STUDENT UNIT GUIDE

NEW EDITION

OCR(A) A2 Chemistry Unit F325
Equilibria, Energetics and Elements

Mike Smith

PHILIP ALLAN

Philip Allan, an imprint of Hodder Education, an Hachette UK company, Market Place, Deddington, Oxfordshire OX15 0SE

Orders

Bookpoint Ltd, 130 Milton Park, Abingdon, Oxfordshire OX14 4SB
tel: 01235 827827
fax: 01235 400401
e-mail: education@bookpoint.co.uk
Lines are open 9.00 a.m.–5.00 p.m., Monday to Saturday, with a 24-hour message answering service. You can also order through the Philip Allan website: www.philipallan.co.uk

ISBN 978-1-4441-7294-2

First printed 2013
Impression number 5 4 3 2 1
Year 2017 2016 2015 2014 2013

Cover photo: Fotolia

Typeset by Integra Software Services Pvt. Ltd., Pondicherry, India

Printed in Dubai

Hachette UK's policy is to use papers that are natural, renewable and recyclable products and made from wood grown in sustainable forests. The logging and manufacturing processes are expected to conform to the environmental regulations of the country of origin.

Contents

Getting the most from this book 4

About this book 5

Content Guidance

Module 1: Rates, equilibrium and pH 8

  How fast? 8

  How far? 16

  Acids, bases and buffers 19

Module 2: Energy 32

  Lattice enthalpy 32

  Entropy 40

  Electrode potentials and fuel cells 43

Module 3: Transition elements 52

  Properties of transition elements 52

Questions & Answers

Q1 The rate equation 65

Q2 Equilibrium 69

Q3 pH 72

Q4 Acids and bases 76

Q5 Born–Haber cycles 79

Q6 Enthalpy, entropy and free energy 85

Q7 Redox equations and electrode potentials 88

Q8 Fuel cells 91

Q9 Transition metal chemistry and redox titrations 93

Q10 Transition metal chemistry 96

Q11 Synoptic question 99

Knowledge check answers 101

Index 103

Getting the most from this book

Questions & Answers

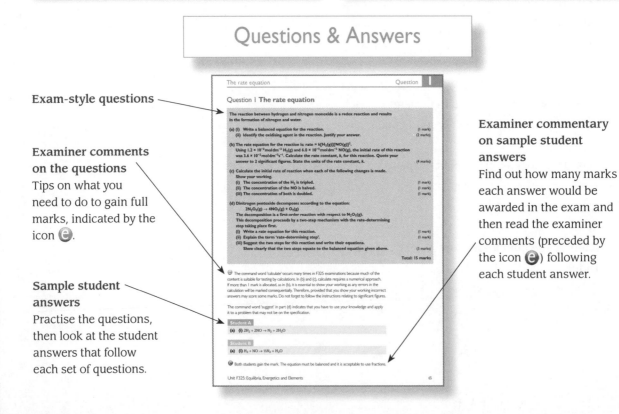

About this book

This unit guide follows on from a series of two AS units and the A2 unit F324 (Rings, Polymers and Analysis). The complete series provides coverage of the whole OCR A-level chemistry specification A. This guide has been written to help you prepare for the Unit F325 exam paper.

The **Content Guidance** section gives a point-by-point description of all the facts you need to know and concepts that you need to understand for Unit F325. It aims to provide you with a basis for your revision. However, you must also be prepared to use other sources in your preparation for the examination.

The **Question and Answer** section shows you the sort of questions you can expect in the unit test. It would be impossible to give examples of every kind of question in one book, but the questions used should give you a flavour of what to expect. Each question has been attempted by two candidates, Student A and Student B. Their answers, along with the examiner's comments, should help you to see what you need to do to score marks — and how you can easily *not* score marks even though you probably understand the chemistry.

What can I assume about the guide?

You can assume that:
- the topics covered in the Content Guidance section relate directly to those in the specification
- the basic facts you need to know are stated clearly
- the major concepts you need to understand are explained
- the questions at the end of the guide are similar in style to those that will appear on the exam paper
- the answers supplied are genuine, combining responses commonly written by students
- the standard of the marking is broadly equivalent to the standard that will be applied to your answers

What can I *not* assume about the guide?

You must *not* assume that:
- every last detail has been covered
- the way in which the concepts are explained is the only way in which they can be presented in an examination (concepts are often presented in an unfamiliar situation)
- the range of question types presented is exhaustive (examiners are always thinking of new ways to test a topic)

Study skills and revision techniques

All students need to develop good study skills and this section provides advice and guidance on how to study A-level chemistry.

Organising your notes

Chemistry students often accumulate a large quantity of notes and it is useful to keep these in a well-ordered and logical manner. It is may be necessary to review your notes on a daily or weekly basis such that the notes you take during lessons are re-written so that they are clear and concise with key points highlighted. It is a good idea to check your notes using textbooks and fill in any gaps. Make sure that you go back and ask the teacher if you are unsure about anything, especially if you find conflicting information in your class notes and textbook.

It is essential to file your notes in specification order using a consistent series of headings. The Content Guidance section can help you with this.

Organising your time

Preparation for examinations is a personal thing. Different people prepare, equally successfully, in different ways. The key is being honest about what works for you.

Whatever your style, you must have a plan. Sitting down the night before the examination with a file full of notes and a textbook does not constitute a revision plan — it is just desperation — and you must not expect a great deal from it. Whatever your personal style, there are a number of things you *must* do and a number of other things you *could* do.

The scheme outlined below is a suggestion as to how you might revise for Unit F325 over 3 weeks. The work pattern is fairly simple.

Day	Week 1	Week 2	Week 3
Mon	Rates of reaction: 20 minutes	Rates of reaction. Write summary notes and attempt questions from books or papers: 30 minutes	The exam paper: 100 marks in 105 minutes. Try a past paper spread over two nights. List anything you are unsure of and ask your teacher for help
Tues	K_c, pH and buffers: 20 minutes Rates of reaction: 10 minutes	K_c, pH and buffers. Write summary notes and attempt questions from books/papers: 30 minutes	
Wed	Lattice energy, ΔH and ΔS: 20 minutes K_c, pH and buffers: 10 minutes Rates of reaction: 5 minutes	Lattice energy, ΔH and ΔS. Write summary notes and attempt questions from books or papers: 30 minutes	Try a second past paper spread over two nights. List anything you are unsure of and ask your teacher for help
Thurs	Electrode potentials and fuel cells: 20 minutes Lattice energy, ΔH and ΔS: 10 minutes K_c, pH and buffers: 5 minutes Rates of reaction: 1 min	Electrode potentials and fuel cells. Write summary notes and attempt questions from books or papers: 30 minutes	

Day	Week 1	Week 2	Week 3
Fri	Transition elements: 20 minutes Electrode potentials and fuel cells: 10 minutes Lattice energy, ΔH and ΔS: 5 minutes K_c, pH and buffers: 1 min	Transition elements. Write summary notes and attempt questions from books or papers: 30 minutes	Re-read (or re-write) all of your summary notes
Sat	Transition elements: 10 minutes Electrode potentials and fuel cells: 5 minutes Lattice energy, ΔH and ΔS: 2 minutes K_c, pH and buffers: 1 min	Get someone to test you on your summary notes: 30 minutes	Try another past paper spread over two nights. List anything you are unsure of and ask your teacher for help
Sun	Spend 5 minutes on each of the topics	Re-write all of your summary notes: 30 minutes	

The revision timetable will probably not suit you and it would be better if you wrote your own to meet your needs. It is only there to give you an idea of how it might work. The most important thing is that the grid at least enables you to see what you should be doing and when you should be doing it. Do not try to be too ambitious — *little and often is by far the best way.*

It might be much more sensible to put together a much longer rolling programme to cover all your A-level subjects, not just chemistry. Try and work out a rolling programme that enables you to cover all your subjects over a 5–6-week period. **Do *not* leave it too late. Start sooner rather than later.**

Content Guidance

Unit F325 builds on a number of different areas in AS chemistry. It requires an understanding of reaction rates and chemical equilibrium from Unit F322. It also extends your understanding of energetics and the calculations associated with it. Throughout this guide the essential pre-knowledge is outlined and references are made to the relevant AS units.

Synoptic assessment

Unit F325 examination papers contain questions that relate to principles first encountered in other units and are designated synoptic questions. These questions relate the content of this module with knowledge and understanding acquired elsewhere in the course. You are expected to apply chemical principles from any part of the entire specification, including:
- mole calculations
- writing balanced equations
- empirical and molecular formula calculations
- bonding and structure

The content of this module is essentially quantitative and relates back to the corresponding qualitative chemistry in Unit F322.

Module 1: Rates, equilibrium and pH

How fast?

Experimental observations show that the rate of a reaction is influenced by temperature, concentration and the use of a catalyst.

The collision theory of reactivity helps to provide explanations for these observations. A reaction cannot take place unless a collision occurs between the reacting particles. Increasing temperature or concentration increases the chance of a collision occurring.

However, not all collisions lead to a successful reaction. The energy of a collision between reacting particles must exceed the minimum energy required to start the reaction. This minimum energy is known as the activation energy, E_a. Increasing the temperature affects the number of collisions with energy that exceeds E_a and the use of a catalyst affects the size of E_a.

Boltzmann distribution of molecular energies

Figure 1 shows a typical distribution of energies at constant temperature.

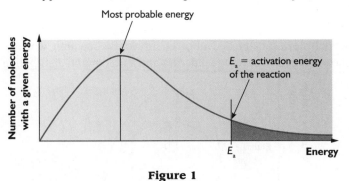

Figure 1

- The area under the curve represents the total number of particles.
- The shaded area represents the number of particles with energy greater than or equal to the activation energy, $E \geq E_a$ (showing the number of particles with sufficient energy to react).

Effect of concentration, temperature and catalyst on the rate of reaction

Concentration

Increasing concentration increases the chance of a collision. The more collisions there are, the faster the reaction will be.

For a gaseous reaction, increasing pressure has the same effect as increasing concentration. When gases react, they react faster at high pressure because there is an increased chance of a collision.

Temperature

An increase in temperature has a dramatic effect on the distribution of energies, as can be seen in Figure 2.

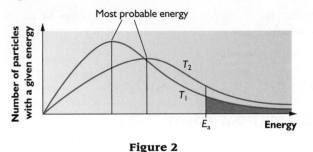

Figure 2

At higher temperatures the distribution flattens and shifts to the right such that:
- there are fewer particles with low energy
- the most probable energy moves to higher energy
- a greater proportion of particles have energy that exceeds the activation energy

Increasing temperature increases the number of particles with energy greater than or equal to the activation energy, $E \geq E_a$, which means that at high temperatures there are more particles with sufficient energy to react and, therefore, the reaction is faster.

Catalyst

Catalysts work by lowering the activation energy for the reaction, which is illustrated below by showing an energy profile diagram and a Boltzmann distribution (Figure 3).

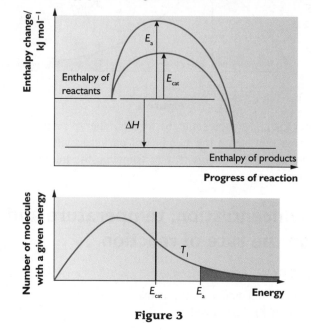

Figure 3

E_a is the activation energy of the uncatalysed reaction and E_{cat} is the activation energy of the catalysed reaction.

At A2, How Fast? builds on your understanding of the AS reaction rates chemistry. It involves measuring and calculating reaction rates using rate equations.

Reaction rate

The rate of a reaction is usually measured as the **change in concentration** of a reaction species **with time**.

The units of rate $= \text{mol dm}^{-3}\,\text{s}^{-1}$

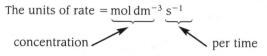

concentration ⟋ ⟍ per time

Measuring rates from graphs

For a reaction $A + B \rightarrow C + D$ it is possible to measure the rate of disappearance of either A or B, or the rate of appearance of one of the products, C or D.

- The rate of *decrease* in concentration of $A = -\dfrac{d[A]}{dt}$

- The rate of *increase* in concentration of $C = \dfrac{d[C]}{dt}$

The gradient of the concentration–time graph is a measure of the rate of a reaction (Figure 4).

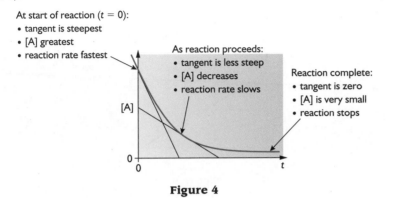

Figure 4

The rate equation

Orders of reaction

The rate equation of a reaction shows how the rate is affected by concentrations of each reactant and can only be determined from experiments.

In general, for a reaction A + B → C + D, the reaction rate is given by:

$$\text{rate} = k[A]^m[B]^n$$

where

- k is the rate constant of the reaction
- m and n are the **orders of reaction** with respect to A and B respectively
- the **overall order** of reaction is $(m + n)$

You know from Unit F322 that increasing concentration usually results in an increased rate of reaction. However, different reagents can behave in a different manner. If we double the concentration of a reagent and the rate increases proportionately (i.e. the rate also doubles) then the reaction is said to be **first order** with respect to that reagent. If by doubling a reagent the reaction increases four-fold, the reaction is said to be **second order** with respect to that reagent, but if increasing the concentration has no effect the reaction is said to be **zero order** with respect to that reagent.

The rate constant

The rate constant, k, indicates the rate of the reaction:

- a large value of k → fast rate of reaction
- a small value of k → slow rate of reaction

An increase in temperature speeds up the rate of most reactions by *increasing* the rate constant, k.

Determination of orders from graphs

During a reaction the concentration of the reagents and product change and it is possible to measure concentration at regular intervals of time. The shape of the resultant graph can be used to predict the order of reaction.

Concentration–time graph

Zero-order reaction

The concentration falls at a steady rate with time (Figure 5).

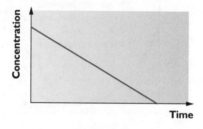

Figure 5

First-order reaction

The concentration halves in equal time periods (Figure 6).

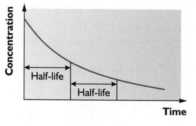

Figure 6

The shape of the graph above indicates the order of the reaction by measuring the **half-life** of a reactant. The **half-life** of a reactant is the time required for its concentration to be reduced by half.

Second-order reaction

The half-life becomes progressively longer as the reaction proceeds. The half-life *increases* with time (Figure 7).

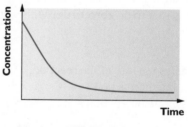

Figure 7

Rate–concentration graphs

A concentration–time graph is first plotted and tangents are drawn at several time values on the concentration–time graph, giving values of reaction rates. A second graph can now be plotted of rate against **concentration**.

Zero-order reaction

The rate is unaffected by changes in concentration (Figure 8).

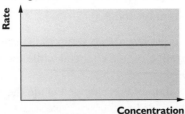

Figure 8

rate $\propto [X]^0$
so, rate = constant

First-order reaction

The rate doubles as the concentration doubles (Figure 9).

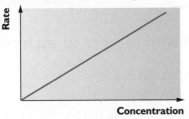

Figure 9

rate $\propto [X]^1$

Second-order reaction

The rate quadruples as the concentration doubles (Figure 10).

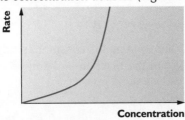

Figure 10

rate $\propto [X]^2$

The second-order relationship can be confirmed by plotting a graph of rate against $[X]^2$, which gives a straight line (Figure 11).

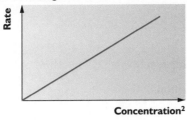

Figure 11

Examiner tip

You are expected to know *two* sets of graphs, the shapes of which can be used to determine the order of reaction. One set plots concentration against time and the other shows the relationship between rate and concentration. It is essential that you do not mix up these graphs.

Measuring rates using the initial rates method

For a reaction, A + B → C + D, carry out experiments using different initial concentrations of the reactants A and B.

Only change one variable at a time so two series of experiments will be required.

- In series 1, the concentration of A is changed while the concentration of B is kept constant.
- In series 2, the concentration of B is changed while the concentration of A is kept constant.

For each experiment plot a concentration–time graph and measure the initial rate from the graph of the tangent drawn at time = 0. A typical set of results looks like:

Experiment	$[A(aq)]/mol\,dm^{-3}$	$[B(aq)]/mol\,dm^{-3}$	Initial rate/$mol\,dm^{-3}\,s^{-1}$
1	1.0×10^{-2}	1.0×10^{-2}	4.0×10^{-3}
2	2.0×10^{-2}	1.0×10^{-2}	1.6×10^{-2}
3	2.0×10^{-2}	2.0×10^{-2}	3.2×10^{-2}

Order of reaction with respect to each reagent

Order of reaction with respect to A

Comparing experiments 1 and 2, [B(aq)] is constant and [A(aq)] is varied. [A(aq)] has doubled, the rate has quadrupled, and the reaction is therefore second order with respect to A(aq), such that:

$$\text{rate} \propto [A(aq)]^2$$

The concentration raised to the power 2 indicates that it is second order with respect to A(aq).

Order of reaction with respect to B

Comparing experiments 2 and 3, [A(aq)] is constant and [B(aq)]is varied. [B(aq)] has doubled, the rate doubles, and the reaction is therefore first order with respect to B(aq), such that:

$$\text{rate} \propto [B(aq)]$$

The rate equation

Combining the two orders with respect to the two reagents gives:

$$\text{rate} \propto [A(aq)]^2[B(aq)]$$

or

$$\text{rate} = k[A]^2[B]$$

and the overall order of this reaction is (2 + 1) = third order.

Rearranging the rate equation:

$$k = \frac{\text{rate}}{[A]^2[B]}$$

and substituting values from experiment 1, gives:

$$k = \frac{4.0 \times 10^{-3}}{(1.0 \times 10^{-2})\,(1.0 \times 10^{-2})}$$

$$= 40\,mol^{-2}\,dm^6\,s^{-1}$$

Units of rate constants

The units of a rate constant depend upon the rate equation for the reaction.

Order	Rate equation	Units of k
First	Rate = $k[A]$	s^{-1}
Second	Rate = $k[A]^2$	$mol^{-1}\,dm^3\,s^{-1}$
Third	Rate = $k[A]^2[B]$ or $k[A]^3$	$mol^{-2}\,dm^6\,s^{-1}$

Knowledge check 3

The table below gives data from the reaction:

$$BrO_3^- + 5Br^- + 6H^+ \rightarrow 3Br_2 + 3H_2O$$

The rate constant, $k = 8.2\,dm^9\,mol^{-3}\,s^{-1}$

[BrO_3^-]/mol dm^{-3}	[Br^-]/mol dm^{-3}	[H^+]/mol dm^{-3}	Initial rate × 10^{-3}/mol dm^3 s^{-1}
0.1	0.2	0.1	1.64
0.2	0.2	0.1	3.28
0.2	0.4	0.1	6.56

Deduce the order of reaction with respect to each reagent.

Determination of reaction mechanisms

The **rate-determining step** is defined as the slowest step in the reaction.

The rate equation can provide clues about a likely reaction mechanism by identifying the slowest stage of a reaction sequence. For instance if the rate equation is:

rate = $k[A]^2[B]$

the slow step will involve **2** mol of A and **1** mol of B, but if the rate equation is:

rate = $k[A][B]^2$

the slow step will involve **1** mol of A and **2** mol of B.

The orders in the rate equation match the number of species involved in the rate-determining step.

Reaction mechanisms often involve many separate steps. You may be asked to use the rate equation and the balanced equation to predict a mechanism that contains two steps. In a two-step mechanism, the rate equation indicates the number of moles of each reactant involved in the **slow** step. The **slow** step + the **fast** step = **balanced** equation.

For the reaction $2H_2(g) + 2NO(g) \rightarrow 2H_2O(l) + N_2(g)$, the rate equation is rate = $k[H_2(g)][NO(g)]^2$. Predict a two-step mechanism.

Knowledge check 1

For the rate equation; rate = $k[A][B]^2$, what will happen to the overall rate of reaction if: (a) [A] is doubled, (b) [B] is halved, (c) both [A] and [B] are trebled?

Knowledge check 2

What are the units of the rate constant, k, for (a) rate = $k[A][B]^2$ and (b) rate = $k[A]^2[B]^2[C]$?

Examiner tip

The slow step in any mechanism is usually the first step. The rate equation tells you the reagents and the number of moles of each reagent involved in the slow step. For example: in a reaction $2A + B + 2C \rightarrow 2X$, if the rate = $k[A][B]^2$ then the slow step involves 1 mol of A, 2 mol of B and no moles of C.

slow step   the rate equation tells us that this involves 1 mol $H_2(g)$ and 2 mol $NO(g)$

  +

fast step

  ‖

balanced equation   the balanced equation is given in the question. Just copy it down.

A possible two-step mechanism is:

slow step $1H_2(g) + 2NO(g) \longrightarrow H_2O(l) + N_2(g) + O(g)$

  +

fast step $1H_2(g) + O(g) \longrightarrow H_2O(l)$

  ‖

balanced equation $2H_2(g) + 2NO(g) \longrightarrow 2H_2O(l) + N_2(g)$

Knowledge check 4

Deduce a two-step mechanism for the reaction $2ICl + H_2 \rightarrow 2HCl + I_2$.

The rate equation is
$r = k[ICl][H_2]$

How far?

This follows on naturally from How fast?. You should look back through the final section on chemical equilibrium in Unit F322 (Chains, Energy and Resources) in this series. This introduces the idea of a reversible reaction, a dynamic equilibrium and le Chatelier's principle.

A **reversible reaction** is a reaction in which the reactants react to form products and the products also react to form the reactants.

A **reversible reaction** is a reaction where the reactants react to form products but the products also re-react to form the reactants. The reaction proceeds in both the forward and the backward directions and leads to the formation of a **dynamic equilibrium**.

A **dynamic equilibrium** is reached when the rate of the forward reaction equals the rate of the reverse reaction so that the composition of the mixture remains constant. However, the reactants and products are interchanging constantly.

Le Chatelier's principle states that if a closed system, at equilibrium, is subjected to a change in temperature, pressure or concentration, the system will change in such a way as to *minimise* the effect of that change.

Temperature

The effect of changing temperature depends on whether or not the forward reaction is exothermic or endothermic.

If the forward reaction is exothermic, $-\Delta H$, increasing temperature will move the position of the equilibrium to the left-hand side.

If the forward reaction is endothermic, $+\Delta H$, increasing temperature will move the position of the equilibrium to the right-hand side.

Pressure

The effect of changing pressure depends on the number of moles of gas on both sides of the equilibrium.

Increasing pressure moves the position of the equilibrium to the side with fewest moles of gas.

Decreasing pressure moves the position of the equilibrium to the side with most moles of gas.

Concentration

The effect of increasing the concentration of either a reagent or a product is to move the position of the equilibrium to the opposite side.

Catalysts

Catalysts speed up the rate of the reaction but do *not* change the position of the equilibrium.

You should be able to use le Chatelier's principle to deduce what happens to equilibrium when it is subjected to stated changes. In the reaction $2NO_2(g) \rightleftharpoons N_2O_4(g)$, $\Delta H = +100\,kJ\,mol^{-1}$.
- If the temperature is increased, the equilibrium moves to the right as it favours the forward reaction because it is endothermic.
- If the pressure is increased, the equilibrium moves to the right because there are fewer moles of gas on the right-hand side of the equilibrium.
- When a catalyst is used, the equilibrium position remains unchanged as a catalyst speeds up the forward and the reverse reactions equally.
- If the concentration of $[NO_2]$ is increased the equilibrium will move to the right to reduce the amount of NO_2 in the equilibrium.

> **Knowledge check 5**
>
> Sulfur trioxide, $SO_3(g)$, is an essential reagent in the production of sulfuric acid. It is obtained from the equilibrium:
>
> $2SO_2(g) + O_2(g) \rightleftharpoons 2SO_3(g)$ $\Delta H = -196\,kJ\,mol^{-1}$
>
> Predict the optimum conditions in terms of temperature and pressure.

The equilibrium constant, K_c

For A2 chemistry, the exact position of equilibrium is calculated using the equilibrium law.

The equilibrium law

K_c is the equilibrium constant in terms of equilibrium concentrations. The equilibrium law states that, for an equation:

$$aA + bB \rightleftharpoons cC + dD$$

$$K_c = \frac{[C]^c[D]^d}{[A]^a[B]^b}$$

- [A], [B], [C] and [D] are the **equilibrium** concentrations of the reactants and products in the reaction.
- Each product and reactant has its equilibrium concentration raised to the **power** of its **balancing number** (*a*, *b* etc.) in the equation.

Working out K_c

For the equilibrium $H_2(g) + I_2(g) \rightleftharpoons 2HI(g)$, applying the equilibrium law gives:

$$K_c = \frac{[HI(g)]^2}{[H_2(g)][I_2(g)]}$$

At equilibrium, $[H_2(g)] = 0.012\,mol\,dm^{-3}$, $[I_2(g)] = 0.001\,mol\,dm^{-3}$, and $[HI(g)] = 0.025\,mol\,dm^{-3}$.

$$K_c = \frac{[HI(g)]^2}{[H_2(g)][I_2(g)]}$$

$$= \frac{0.025^2}{0.012 \times 0.001} = 52.1$$

Units of K_c

The units of K_c depend upon the equilibrium expression for the reaction. Each concentration value is replaced by its units:

$$K_c = \frac{[HI(g)]^2}{[H_2(g)][I_2(g)]} = \frac{(mol\,dm^{-3})^2}{(mol\,dm^{-3})(mol\,dm^{-3})}$$

For this equilibrium, the units cancel and K_c has no units.

Writing expressions for K_c

It is essential that you are able to write expressions for K_c and are able to deduce the units, if any, for each expression.

Equilibrium

	$2NO_2(g) \rightleftharpoons N_2O_4(g)$	$N_2(g) + 3H_2(g) \rightleftharpoons 2NH_3(g)$	$Br_2(g) + H_2(g) \rightleftharpoons 2HBr(g)$
K_c	$\dfrac{[N_2O_4\,g)]}{[NO_2(g)]^2}$	$\dfrac{[NH_3(g)]^2}{[N_2(g)][H_2(g)]^3}$	$\dfrac{[HBr(g)]^2}{[H_2(g)][Br_2(g)]}$
Units	$mol^{-1}\,dm^3$	$mol^{-2}\,dm^6$	None

Properties of K_c

K_c indicates how *far* a reaction proceeds but tells us *nothing* about how *fast* the reaction occurs. The size of K_c indicates the extent of a chemical equilibrium.

If K_c is large (K_c = 1000) the equilibrium lies to the right-hand side and there will be a high percentage of product formed.

If K_c is small (K_c = 1 × 10^{-3}) the equilibrium lies to the left-hand side and there will be a low percentage of product formed.

If K_c = 1 the equilibrium lies halfway between reactants and products.

Changing K_c

K_c is a constant *but* it is temperature dependent. K_c is unaffected by changes in concentration or pressure but K_c can be changed by altering the temperature.

It is easy to see why changing a concentration does not change K_c but more difficult when considering a change in pressure.

In the Haber process: $N_2(g) + 3H_2(g) \rightleftharpoons 2NH_3(g)$, le Chatelier's principle predicts that an increase in pressure moves the equilibrium to the right, which is true. If nothing else happened the value of K_c would increase. However, increasing pressure also decreases the volume which changes the concentrations. This change in concentrations ensures that K_c remains constant.

In an **exothermic** reaction K_c *decreases* with increasing temperature because raising the temperature reduces the equilibrium yield of products.

In an **endothermic** reaction, K_c *increases* with increasing temperature because raising the temperature increases the equilibrium yield of products.

Knowledge check 6

What are the units of K_c for (a) $PCl_3(g) + Cl_2(g) \rightleftharpoons PCl_5(g)$ and (b) $2SO_2(g) + O_2(g) \rightleftharpoons 2SO_3(g)$?

Examiner tip

When calculating a value for K_c, first check the units. If there are no units it does not matter whether you use moles or concentrations of each chemical. For a reaction such as $2NO_2(g) \rightleftharpoons N_2O_4(g)$ it is necessary to convert the moles of gas into concentrations before calculating K_c.

Knowledge check 7

$N_2(g)$ and $H_2(g)$ were mixed in a 4.0 dm^3 container and allowed to reach equilibrium. The equilibrium mixture contained 5.0 mol $N_2(g)$, 10.0 mol $H_2(g)$ and 5.0 mol $NH_3(g)$. Calculate K_c for $N_2(g) + 3H_2(g) \rightleftharpoons 2NH_3(g)$.

Determination of K_c from experiment

The equilibrium constant, K_c, can be determined from experimental results.

Example

This example illustrates how to answer a typical question.

0.200 mol CH_3COOH and 0.100 mol C_2H_5OH were mixed together with a trace of acid catalyst in a total volume of 200 cm³. The mixture was allowed to reach equilibrium:

$$CH_3COOH + C_2H_5OH \rightleftharpoons CH_3COOC_2H_5 + H_2O$$

Analysis of the mixture showed that 0.115 mol of CH_3COOH were present at equilibrium. Calculate the equilibrium constant, K_c.

From the information given, the number of moles of CH_3COOH that reacted = 0.200 − 0.115 = 0.085. The balanced equation tells us the molar ratio of the reactants and the products:

Balanced equation	CH_3COOH +	C_2H_5OH	$\rightleftharpoons$	$CH_3COOC_2H_5$ +	H_2O
Molar ratio	1 mol	1 mol	$\longrightarrow$	1 mol	1 mol
Change/mol	−0.085	−0.085		+0.085	+0.085

	CH_3COOH +	C_2H_5OH	$\rightleftharpoons$ $CH_3COOC_2H_5$ +	H_2O
Initial amount/mol	0.200	0.100	0	0
Change in moles	−0.085	−0.085	+0.085	+0.085
Equilibrium amount/mol	0.115	0.015	0.085	0.085
Equilibrium concentration / mol dm⁻³	0.115/0.20	0.015/0.20	0.085/0.20	0.085/0.20

Write the expression for K_c, substitute values and calculate K_c:

$$K_c = \frac{[CH_3COOC_2H_5]\,[H_2O]}{[CH_3COOH]\,[C_2H_5OH]}$$

$$= \frac{\left(\dfrac{0.085}{0.20}\right) \times \left(\dfrac{0.085}{0.20}\right)}{\left(\dfrac{0.115}{0.20}\right) \times \left(\dfrac{0.015}{0.20}\right)}$$

$$= \frac{0.425 \times 0.425}{0.575 \times 0.075} = 4.19$$

The answer does not have any units because they all cancel.

Acids, bases and buffers

Acids were introduced in both AS units. You should be able to define an acid as a proton donor and be able to write equations for the reactions of acids.

> **Knowledge check 8**
>
> 0.4 mol $H_2(g)$ and 0.3 mol $I_2(g)$ were mixed in a 2 dm³ flask and allowed to reach equilibrium. The equilibrium mixture contained 0.2 mol HI(g) Deduce the amount of H_2 and I_2 in the equilibrium mixture and calculate K_c for $H_2(g) + I_2(g) \rightleftharpoons 2HI(g)$.

Brønsted–Lowry acids and bases

An acid–base reaction involves proton transfer.

$$NaOH(aq) + HCl(aq) \rightarrow NaCl(aq) + H_2O(l)$$

which can be simplified to:

$$H^+(aq) + OH^-(aq) \rightarrow H_2O(l)$$

Acids also react with carbonates:

$$Na_2CO_3(aq) + 2HCl(aq) \rightarrow 2NaCl(aq) + CO_2(g) + H_2O(l)$$

giving the ionic equation

$$CO_3^{2-}(aq) + 2H^+(aq) \rightarrow CO_2(g) + H_2O(l)$$

In this case the acid donates protons to the carbonate, which splits into carbon dioxide and water. The carbonate is therefore a base in the reaction.

Knowledge check 9

Deduce the formula of the salt formed from each of the following acid–base reactions:
(a) $CH_3COOH + KOH$
(b) $HCOOH + Mg(OH)_2$
(c) $H_3PO_4 + CaCO_3$

Acid–base pairs

A molecule of an acid contains a hydrogen that can be released as a positive hydrogen ion or proton, H^+.

An acid–base **conjugate pair** are linked together by H^+; the **conjugate acid** donates H^+ and the **conjugate base** accepts H^+.

Acids and bases are linked by H^+ as **conjugate pairs** such that the **conjugate acid** donates H^+ and the **conjugate base** accepts H^+. An acid can only donate a proton if there is a base to accept it. By mixing an acid with a base, an equilibrium is set up between *two* acid–base **conjugate pairs** (Figure 12).

In the forward reaction: ⟶

- $CH_3COOH(aq)$ donates a H^+ to the water and, therefore, behaves as an acid
- H_2O accepts a H^+ from $CH_3COOH(aq)$ and, therefore, behaves as a base

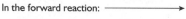

$$CH_3COOH(aq) \ + \ H_2O(l) \ \rightleftharpoons \ H_3O^+ \ + \ CH_3COO^-(aq)$$

⟵ In the reverse reaction:

- H_3O^+ donates a H^+ to the $CH_3COO^-(aq)$ and, therefore, behaves as an acid
- $CH_3COO^-(aq)$ accepts a H^+ from H_3O^+ and, therefore, behaves as a base

Figure 12

$CH_3COOH(aq)$ and $CH_3COO^-(aq)$ form an acid–base conjugate pair and H_3O^+ and $H_2O(l)$ form a second acid–base conjugate pair.

In the equilibrium $NH_3(g) + H_2O(l) \rightleftharpoons NH_4^+(aq) + OH^-(aq)$ it can be seen that, in the forward reaction, the water has donated a proton to the ammonia and is therefore the acid while the ammonia is the base. For the reverse reaction the ammonium ion is the acid and the hydroxide is the base.

Knowledge check 10

Identify the conjugate **base** for each of the following: (a) H_2O
(b) HSO_4^- (c) NH_3
(d) C_6H_5COOH.

In summary:

Knowledge check 11

Identify the conjugate **acid** for each of the following:
(a) H_2O (b) HSO_4^-
(c) NH_3 (d) CH_3NH_2.

$$NH_3(g) \ + \ H_2O(l) \ \rightleftharpoons \ NH_4^+(aq) \ + \ OH^-(aq)$$

Base 2　　　**Acid 1**　　　**Acid 2**　　　**Base 1**

Strengths of acids and bases

The acid–base equilibrium of an acid, HA, in water is:

$$HA(aq) + H_2O(l) \rightleftharpoons H_3O^+(aq) + A^-(aq)$$

or, to emphasise the loss of a proton, H^+, by dissociation:

$$HA(aq) \rightleftharpoons H^+(aq) + A^-(aq)$$

The strength of an acid shows the extent of dissociation into H^+ and A^-.

Strong acids

Acids vary considerably in the ease with which they are able to release their hydrogen ions. A strong acid, such as nitric acid, HNO_3, is a good proton donor with the equilibrium position well to the right.

$$\overset{\text{equilibrium} \longrightarrow}{HNO_3(aq) \rightleftharpoons H^+(aq) + NO_3^-(aq)}$$

There is almost complete dissociation and it is usual to write the equation as:

$$HNO_3(aq) \rightarrow H^+(aq) + NO_3^-(aq)$$

Weak acids

A weak acid, such as ethanoic acid, CH_3COOH, is a poor proton donor with the equilibrium position well to the left.

$$\overset{\longleftarrow \text{equilibrium}}{CH_3COOH(aq) \rightleftharpoons H^+(aq) + CH_3COO^-(aq)}$$

There is only partial dissociation.

It is important to distinguish between the terms 'strong' and 'concentrated'.

A **strong** acid is one that is highly ionised in aqueous solution. A **concentrated** acid is one made by dissolving large amounts of the acid in a small volume of water.

A **weak** acid is one that is only partially ionised in aqueous solution. A **dilute** acid is one made by dissolving small amounts of the acid in a large volume of water.

A similar distinction is made for bases, but you meet few weak bases in the A2 course. Amines and ammonia are weak bases, as they are only ionised to a small extent, but all the metal hydroxides are strong bases. Again it is an easy mistake to make to use the word 'weak' when you mean 'dilute'. Limewater (aqueous calcium hydroxide) is, for example, a strong base but, as it has a low solubility it is inevitably always dilute.

The usual way of indicating the strength of an acid is to use the equilibrium constant for its ionisation in water.

The acid dissociation constant, K_a

The extent of acid dissociation is shown by an equilibrium constant called the acid dissociation constant, K_a.

Examiner tip

To illustrate acid–base conjugate pairs, questions may use two acids — one strong and one weak. Remember that the stronger acid (the acid with the lower pH or pK_a value) will donate a proton to the weaker acid.

Knowledge check 12

Ethanoic acid is mixed with nitric acid forming an equilibrium containing acid–base conjugate pairs. Complete the following equation:

$CH_3CO_2H + HNO_3 \rightleftharpoons$

.... +

....

For the reaction $HA(aq) \rightleftharpoons H^+(aq) + A^-(aq)$

$$K_a = \frac{[H^+(aq)][A^-(aq)]}{[HA(aq)]}$$

The units are:

$$K_a = \frac{[(mol\ dm^{-3})^2]}{(mol\ dm^{-3})} = mol\ dm^{-3}$$

A **large K_a** value shows that the extent of dissociation is large and the acid is a strong acid.

A **small K_a** value shows that the extent of dissociation is small and the acid is a weak acid.

Ethanoic acid ionises as:

$$CH_3COOH(aq) \rightleftharpoons CH_3COO^-(aq) + H^+(aq)$$

and, as with the other equilibria, an equilibrium constant can be defined for this reaction as:

$$K_a = \frac{[CH_3COO^-(aq)][H^+(aq)]}{[CH_3COOH(aq)]}$$

To indicate that the reaction involves an acid it is conventional to provide the equilibrium constant K with the subscript 'a'. K_a is called the **acid dissociation constant**.

For ethanoic acid this has a value of 1.7×10^{-5}, which makes it clear that a solution of ethanoic acid consists largely of ethanoic acid molecules with only relatively few ethanoate ions and hydrogen ions.

Methanoic acid has a K_a value of 1.6×10^{-4}, which is almost ten times larger than the figure for ethanoic acid. This tells us that methanoic acid, though weak, is stronger than ethanoic acid.

The mineral acids have much larger values for K_a. Nitric acid, for example, is approximately 40 and sulfuric acid is often listed with just the comment 'very large'.

Calculating hydrogen ion concentrations

The pH scale

pH is defined by the equation: pH = $-\log_{10}$ [H$^+$(aq)].

The concentrations of $H^+(aq)$ ions in acid solutions vary widely between about $10\ mol\ dm^{-3}$ and about $1 \times 10^{-15}\ mol\ dm^{-3}$. The pH scale (Figure 13) is used to overcome the problem of these small numbers. This is a logarithmic scale, each change of 1 on the pH scale corresponds to a ten-fold change in the [H$^+$(aq)].

pH	0	1	2	3	4	5	6	7	8	9	10	11	12	13	14
[H$^+$]	1	10^{-1}	10^{-2}	10^{-3}	10^{-4}	10^{-5}	10^{-6}	10^{-7}	10^{-8}	10^{-9}	10^{-10}	10^{-11}	10^{-12}	10^{-13}	10^{-14}

More acidic ← —————— Neutral —————— → More alkaline

Figure 13

You should be able to convert between pH and $H^+(aq)$ and vice versa using the relationships below:

$$pH = -\log_{10}[H^+(aq)]$$

and

$$[H^+(aq)] = 10^{-pH}$$

Calculating the pH of strong acids

For a strong acid, we can assume **complete dissociation** and the concentration of $H^+(aq)$ can be found from the acid concentration.

Example 1

A strong acid, HA, has a concentration of $0.020 \, mol \, dm^{-3}$. What is the pH?

There is complete dissociation, therefore $[H^+(aq)] = 0.020 \, mol \, dm^{-3}$, and:

$$pH = -\log_{10}[H^+(aq)]$$

$$= -\log_{10} 0.020 = 1.7$$

Example 2

A strong acid, HA, has a pH of 2.4. What is the concentration of $H^+(aq)$?

There is complete dissociation, therefore $[H^+(aq)] = 10^{-pH} = 10^{-2.4} \, mol \, dm^{-3}$, and:

$$[H^+(aq)] = 3.98 \times 10^{-3} \, mol \, dm^{-3}$$

Knowledge check 13

Calculate the pH of
(a) $0.1 \, mol \, dm^{-3}$ HCl(aq),
(b) $0.01 \, mol \, dm^{-3}$ HCl(aq),
(c) $0.001 \, mol \, dm^{-3}$ HCl(aq).

Calculating the pH of weak acids

To calculate the pH of a strong acid all that you need to know is the concentration of the strong acid. Weak acids do not completely dissociate and to calculate the pH of a weak acid, HA, you need to know:
- the concentration of the acid
- the acid dissociation constant, K_a

In the equilibrium of a weak aqueous acid, $HA(aq) \rightleftharpoons H^+(aq) + A^-(aq)$, we assume that:
- only a very small proportion of HA dissociates and hence the amount of undissociated acid is the same as the initial concentration of $[HA(aq)]$
- there is a negligible proportion of $H^+(aq)$ from ionisation of water, such that $[H^+(aq)] = [A^-(aq)]$

Using these approximations:

$$K_a = \frac{[H^+(aq)][A^-(aq)]}{[HA(aq)]} \approx \frac{[H^+(aq)]^2}{[HA(aq)]}$$

Example

For a weak acid, $[HA(aq)] = 0.200 \, mol \, dm^{-3}$, and $K_a = 1.70 \times 10^{-4} \, mol \, dm^{-3}$ at 25°C. Calculate the pH:

$$K_a = \frac{[H^+(aq)][A^-(aq)]}{[HA(aq)]} \approx \frac{[H^+(aq)]^2}{[HA(aq)]}$$

Therefore:

$$1.70 \times 10^{-4} = \frac{[H^+(aq)]^2}{0.200}$$

$$(1.70 \times 10^{-4}) \times 0.200 = [H^+(aq)]^2$$

$$[H^+(aq)] = \sqrt{(1.70 \times 10^{-4}) \times 0.200}$$

$$= 5.83 \times 10^{-3} = 0.00583$$

$$pH = \log_{10}[H^+(aq)]$$

$$= -\log_{10} 0.00583 = 2.23$$

An alternative way of doing this type of calculation is to use the equation:

$$pH = -\log_{10} \sqrt{(K_a \text{ of } HA) \times (\text{concentration of } HA)}$$

or

$$pH = -\log_{10}\sqrt{K_a \times [HA]}$$

The same calculation can now be carried out in a single step:

$$pH = -\log_{10}\sqrt{1.70 \times 10^{-4} \times 0.200}$$

$$= 2.23$$

K_a and pK_a

As with $[H^+(aq)]$ and pH, K_a is often expressed in a logarithmic form, pK_a, which is defined as:

$$pK_a = -\log_{10} K_a$$

With this logarithmic scale, each change of 1 on the pK_a scale corresponds to a tenfold change in the K_a. Like pH, pK_a can be used as a guide to the acidity. The lower the pK_a value the stronger the acid.

Acid		K_a/mol dm^{-3}	pK_a
Ethanoic acid	CH_3COOH	1.7×10^{-5}	$-\log_{10}(1.7 \times 10^{-5}) = 4.77$
Benzoic acid	C_6H_5COOH	6.3×10^{-5}	$-\log_{10}(6.3 \times 10^{-5}) = 4.20$

This indicates that benzoic acid is a stronger acid than ethanoic acid.

Knowledge check 14

The K_a values of three acids A, B and C are A = $7.70 \times 10^{-4} \, mol \, dm^{-3}$, B = $6.60 \times 10^{-3} \, mol \, dm^{-3}$ and C = $4.20 \times 10^{-4} \, mol \, dm^{-3}$. Put the acids in order of strength starting with the strongest acid.

pK_a is defined by the equation $pK_a = -\log_{10}(K_a)$.

Examiner tip

When carrying out calculations it is important not to round numbers until the end of the calculation. Make sure you know how to store numbers in your calculator.

Knowledge check 15

Use the data in the table to calculate how much stronger benzoic acid is than ethanoic acid.

The ionic product of water, K_w

Water ionises very slightly, acting as both an acid and a base:

$H_2O(l)$	+	$H_2O(l)$	$\rightleftharpoons$	$H_3O^+(aq)$	+	$OH^-(aq)$
Acid 1		Base 2		Acid 2		Base 1
(donates proton)		(accepts proton)		(donates proton)		(accepts proton)

or, more simply,

$$\overleftarrow{\text{equilibrium}}$$

$$H_2O(l) \rightleftharpoons H^+(aq)\ OH^-(aq)$$

In water, a very small proportion of molecules dissociates into $H^+(aq)$ and $OH^-(aq)$ ions. Treating water as a weak acid:

$$K_a = \frac{[H^+(aq)][OH^-(aq)]}{[H_2O(l)]}$$

Rearranging gives:

$$\underbrace{K_a \times [H_2O(l)]}_{\text{constant, } K_w} = [H^+(aq)][OH^-(aq)]$$

K_w is called the **ionic product** of water:

$$K_w = [H^+(aq)][OH^-(aq)] = 1.0 \times 10^{-14}\ \text{mol}^2\,\text{dm}^{-6}\ \text{(at 25°C)}.$$

K_w is temperature dependent and is only equal to $1.0 \times 10^{-14}\ \text{mol}^2\,\text{dm}^{-6}$ at 25°C (298 K).

At 10°C (283 K), $K_w = 2.9 \times 10^{-15}\ \text{mol}^2\,\text{dm}^{-6}$, but at 40°C (313 K), $K_w = 2.9 \times 10^{-14}\ \text{mol}^2\,\text{dm}^{-6}$.

Using K_w to calculate the pH of water

At 25°C $K_w = [H^+(aq)][OH^-(aq)] = 1.0 \times 10^{-14}\ \text{mol}^2\,\text{dm}^{-6}$. Assume that $[H^+(aq)] = [OH^-(aq)]$, such that $K_w = [H^+(aq)]^2 = 1.0 \times 10^{-14}\ \text{mol}^2\,\text{dm}^{-6}$:

$$[H^+(aq)] = 1.0 \times 10^{-7}\ \text{mol}^2\,\text{dm}^{-6}$$

$$pH = -\log_{10}[H^+(aq)] = -\log_{10}(1.0 \times 10^{-7}) = 7$$

However, since K_w changes with temperature it follows that the pH of water is only equal to 7 at 25°C.

At a temperature below 25°C, $K_w = [H^+(aq)][OH^-(aq)] = 2.9 \times 10^{-15}\ \text{mol}^2\,\text{dm}^{-6}$. Assume that $[H^+(aq)] = [OH^-(aq)]$ such that $K_w = [H^+(aq)]^2 = 2.9 \times 10^{-15}\ \text{mol}^2\,\text{dm}^{-6}$:

$$[H^+(aq)] = 5.4 \times 10^{-8}\ \text{mol}^2\,\text{dm}^{-6}$$

$$pH = -\log_{10}[H^+(aq)] = -\log_{10}(5.4 \times 10^{-8}) = 7.3$$

At a temperature above 25°C, $K_w = [H^+(aq)][OH^-(aq)] = 2.9 \times 10^{-14}\ \text{mol}^2\,\text{dm}^{-6}$. Assume that $[H^+(aq)] = [OH^-(aq)]$, such that $K_w = [H^+(aq)]^2 = 2.9 \times 10^{-14}\ \text{mol}^2\,\text{dm}^{-6}$:

$$[H^+(aq)] = 1.7 \times 10^{-7}\ \text{mol}^2\,\text{dm}^{-6}$$

$$pH = -\log_{10}[H^+(aq)] = -\log_{10}(1.7 \times 10^{-7}) = 6.8$$

> **Knowledge check 16**
>
> The ionisation product, K_w, for $H_2O(l) \rightleftharpoons H^+(aq) + OH^-(aq)$ is temperature dependent. At 25°C, $K_w = 1.0 \times 10^{-14}\ \text{mol}\,\text{dm}^{-3}$; at 50°C, $K_w = 5.6 \times 10^{-4}\ \text{mol}\,\text{dm}^{-3}$. Deduce the sign of ΔH for $H_2O(l) \rightleftharpoons H^+(aq) + OH^-(aq)$.

Knowledge check 17

Calculate the pH of each of the following aqueous solutions:

(a) $0.15\,mol\,dm^{-3}$ HNO_3

(b) $0.15\,mol\,dm^{-3}$ HCN ($K_a = 4.8 \times 10^{-10}\,mol\,dm^{-3}$)

(c) $0.15\,mol\,dm^{-3}$ $NaOH$

(d) a mixture of $20\,cm^3$ of $1.0\,mol\,dm^{-3}$ HCl and $10\,cm^3$ of $1.0\,mol\,dm^{-3}$ $NaOH$

Using K_w to calculate the pH of strong alkalis

The pH of a strong alkali, such as NaOH, can be calculated from the concentration of the alkali and the ionic product of water, K_w.

Example

A strong alkali, KOH, has a concentration of $0.50\,mol\,dm^{-3}$. What is the pH at 25°C?

$KOH(aq) \rightarrow K^+(aq) + OH^-(aq)$ and since the dissociation is complete, $[OH^-(aq)] = [KOH(aq)] = 0.50\,mol\,dm^{-3}$. First find $[H^+]$ using K_w and $[OH^-]$:

$$K_w = [H^+(aq)][OH^-(aq)] = 1 \times 10^{-14}\,mol^2\,dm^{-6}$$

Therefore:

$$[H^+(aq)] = \frac{K_w}{[OH^-(aq)]} = \frac{1 \times 10^{-14}}{0.50} = 2 \times 10^{-13}\,mol\,dm^{-3}$$

$$pH = -\log_{10}[H^+(aq)] = -\log_{10}(2 \times 10^{-13}) = 12.7$$

Buffer solutions

A **buffer solution** is a solution that resists changes in pH during the addition of an acid or an alkali and maintains a near constant pH by removing most of the added acid or alkali.

A **buffer solution** is a solution that 'resists' changes in pH during the addition of small amounts of an acid or an alkali.

Making a buffer solution

A buffer solution is a mixture of a weak acid, HA, and its conjugate base, A^-:

$$HA(aq) \quad \rightleftharpoons \quad H^+(aq) \quad + \quad A^-(aq)$$

weak acid   conjugate base

An example of a common buffer solution is a mixture of CH_3COOH (the weak acid) and $CH_3COO^-Na^+$ (the conjugate base) (Figure 14). A mixture of any weak acid and the salt of that weak acid can be used as a buffer. The pH at which the buffer operates depends on the K_a of the weak acid and the relative concentrations of the weak acid and the conjugate base.

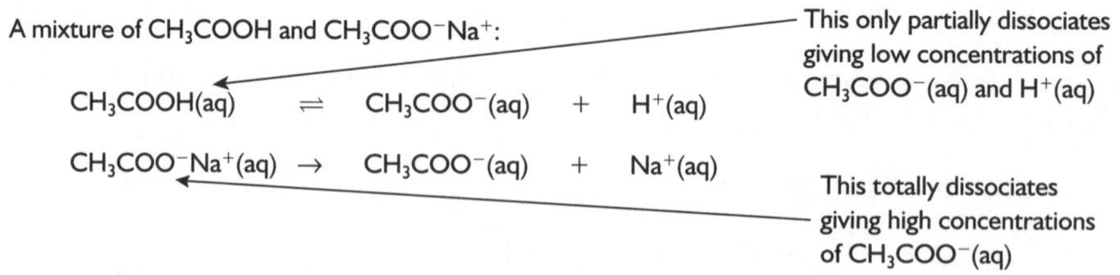

Examiner tip

Many buffers are prepared by mixing together a weak acid and a strong base — the weak acid *must* be in excess. For example, when $100\,cm^3$ of $0.2\,mol\,dm^{-3}$ $CH_3COOH(aq)$ is mixed with $100\,cm^3$ of $0.1\,mol\,dm^{-3}$ $NaOH(aq)$, the resultant mixture contains unreacted $CH_3COOH(aq)$ and its salt $CH_3COO^-Na^+(aq)$ which are the essential components of the buffer.

A mixture of CH_3COOH and $CH_3COO^-Na^+$:

$CH_3COOH(aq) \quad \rightleftharpoons \quad CH_3COO^-(aq) \quad + \quad H^+(aq)$ — This only partially dissociates giving low concentrations of $CH_3COO^-(aq)$ and $H^+(aq)$

$CH_3COO^-Na^+(aq) \rightarrow \quad CH_3COO^-(aq) \quad + \quad Na^+(aq)$ — This totally dissociates giving high concentrations of $CH_3COO^-(aq)$

Figure 14

The high $[CH_3COO^-(aq)]$ forces the equilibrium back to the left-hand side and results in the buffer solution containing very low $[H^+(aq)]$ and very high $[CH_3COOH(aq)]$ and very high $[CH_3COO^-(aq)]$. The high concentrations of $[CH_3COOH(aq)]$ and $[CH_3COO^-(aq)]$ resist any changes in pH.

How does a buffer act?

A buffer solution contains three important components:
- high concentration of the weak acid [$CH_3COOH(aq)$]
- high concentration of the conjugate base [$CH_3COO^-(aq)$]
- low concentration of [$H^+(aq)$]

On addition of an acid, [$H^+(aq)$], the high concentration of the conjugate base, $CH_3COO^-(aq)$, removes most of the added [$H^+(aq)$] by forming $CH_3COOH(aq)$:

$$CH_3COO^-(aq) + H^+(aq) \rightarrow CH_3COOH(aq)$$

On addition of an alkali, [$OH^-(aq)$], the high concentration of $CH_3COOH(aq)$ removes most of the added [$OH^-(aq)$] by forming $CH_3COO^-(aq)$:

$$CH_3COOH(aq) + OH^-(aq) \rightarrow H_2O(l) + CH_3COO^-(aq)$$

A buffer cannot cancel out the effect of any acid or alkali that is added. The buffer removes most of any acid or alkali that is added and *minimises* any changes in pH.

Calculations involving buffer solutions

The pH of a buffer solution depends upon the acid dissociation constant, K_a, of the buffer system and the ratio of the weak acid and its conjugate base.

For a buffer containing the weak acid, HA, and its conjugate base, the salt A⁻:

$$K_a = \frac{[salt][H^+]}{[acid]} \qquad \frac{K_a[acid]}{[salt]} = [H^+]$$

Remember that pH = $-\log_{10}[H^+]$. $\therefore$ pH = $-\log_{10}\dfrac{K_a[acid]}{[salt]}$

It follows that the pH of a buffer can be altered by adjusting the weak acid/conjugate base ratio.

Example

(a) Calculate the pH of a buffer with concentrations of $0.10\,mol\,dm^{-3}$ $CH_3COOH(aq)$ and $0.10\,mol\,dm^{-3}$ $CH_3COO^-(aq)$.

(b) What happens to the pH if the concentration of $CH_3COOH(aq)$ is changed to $0.30\,mol\,dm^{-3}$ CH_3COOH; $K_a = 1.7 \times 10^{-5}\,mol\,dm^{-3}$.

(a) First calculate [$H^+(aq)$] using:

$$[H^+(aq)] = K_a \times \frac{[HA(aq)]}{[A^-(aq)]}$$

$$[H^+(aq)] = 1.7 \times 10^{-5} \times \frac{0.10}{0.10}$$

$$[H^+(aq)] = 1.7 \times 10^{-5}\,mol\,dm^{-3}$$

Then calculate pH from [$H^+(aq)$]

$\therefore$ pH = $-\log_{10}[H^+(aq)]$
$= -\log_{10}(1.7 \times 10^{-5})$

$\therefore$ pH of the buffer solution = 4.77

(b) First calculate [$H^+(aq)$] using:

$$[H^+(aq)] = K_a \times \frac{[HA(aq)]}{[A^-(aq)]}$$

$$[H^+(aq)] = 1.7 \times 10^{-5} \times \frac{0.30}{0.10}$$

$$[H^+(aq)] = 5.1 \times 10^{-5}\,mol\,dm^{-3}$$

Then calculate pH from [$H^+(aq)$]

$\therefore$ pH = $-\log_{10}[H^+(aq)]$
$= -\log_{10}(5.1 \times 10^{-5})$

$\therefore$ pH of the buffer solution = 4.29

Control of pH in blood

In the human body the blood plasma has a normal pH of 7.4. If the pH falls below 7.0 or rises above 7.8 the results could be fatal. The buffer systems in the blood are extremely effective and protect the fluid from large changes in pH. Blood contains a number of buffering systems, the major one being the carbonic acid–hydrogen carbonate (bicarbonate) system:

$$H_2O(l) \quad + \quad CO_2(g) \quad \rightleftharpoons \quad H_2CO_3(aq) \quad \rightleftharpoons \quad HCO_3^-(aq) \quad + \quad H^+(aq)$$

Carbonic acid · Hydrogencarbonate

Adding an acid to the system will increase the concentration of $H^+(aq)$, driving the equilibrium to the left-hand side. This increases the concentration of the carbonic acid, H_2CO_3, which in turn is decreased by an increased rate of breathing such that more $CO_2(g)$ is exhaled resulting in the H_2CO_3 moving further to the left to replace it. The two equilibria together resist the increase in acidity.

Adding an alkali to the system will decrease the concentration of $H^+(aq)$, driving the equilibrium to the right-hand side. This decreases the concentration of the carbonic acid, H_2CO_3, which in turn is increased by a decreased rate of breathing such that less $CO_2(g)$ is exhaled resulting in the H_2CO_3 being replaced. The two equilibria together resist the increase in basicity.

Common everyday uses of buffers

Soaps and detergents are alkaline and irritate the skin and eyes. Shampoo often contains a mixture of citric acid (a weak acid) and sodium citrate (the conjugate base) which acts as a buffer. The ratio of the acid and the conjugate base is adjusted so that the pH is maintained at about pH 5.5, which is approximately the pH of skin.

Babies often suffer from nappy rash because dirty nappies contain ammonia, which is a weak base and has a pH in the region 7 to 9. Bacteria multiply rapidly in the pH region 7 to 9 but not at all at pH 6. Baby lotions are buffered around 5.5 to 6.0, the approximate pH of skin, such that the lotion protects the baby but kills the bacteria.

pH changes and indicators

An **indicator** is a substance that changes colour with a change in pH.

Many **indicators** are weak acids and can be represented as HIn. The weak acid, HIn, and its conjugate base, In⁻, have different colours; e.g. for methyl orange:

	RED				YELLOW
	HIn(aq)	$\rightleftharpoons$	$H^+(aq)$	+	In⁻(aq)
	Weak acid				Conjugate base

At the end point of a titration HIn and In⁻ are present in equal concentrations.

Using methyl orange as indicator:
- at the end point [HIn] (red) = [In⁻] (yellow)
- the colour at the end point is orange from equal proportions of red ([HIn]) and yellow ([In⁻])
- the pH of the end point is called the pK_{In} of the indicator

Knowledge check 18

The indicator bromocresol green, $K_a = 2 \times 10^{-5}\,mol\,dm^{-3}$, is yellow at low pH and turns blue at high pH. Calculate the pH at which the indicator turns green.

pH ranges for common indicators

An indicator changes colour over a range of about two pH units within which is the pK_{In} value of the indicator (Figure 15).

pH   0   1   2   3   4   5   6   7   8   9   10   11   12   13   14

RED ⟵————⟶ YELLOW
methyl orange, $pK_{In} = 3.7$

COLOURLESS ⟵————⟶ PINK
phenolphthalein, $pK_{In} = 9.3$

Figure 15

Choosing an indicator

When the acid and the base have completely reacted this is known as the **equivalence point**. At the equivalence point of the titration there is a sharp change in pH for a very small addition of acid or base.

The choice of a suitable indicator is best shown using titration curves.

Plotting titration curves

A titration curve shows the changes in pH during a titration.

Key features of titration curves

The pH changes rapidly at the near vertical portion of the titration curve. This is the **end point** of the titration.

The sharp change in pH is brought about by a very small addition of alkali, typically the addition of one drop.

The indicator is only suitable if its pK_{In} value is within the pH range of the near vertical portion of the titration curve.

Choosing an indicator using titration curves

On the titration curves below (Figure 16):
- different combinations of strong and weak acids have been used
- the pK_{In} values are shown for the indicators methyl orange (**MeO**) and phenolphthalein (**Ph**)

Strong acid/strong alkali

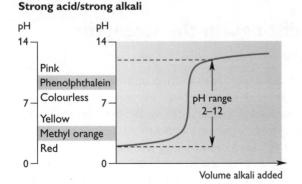

Methyl orange (**MeO**) ✓
Phenolphthalein (**Ph**) ✓

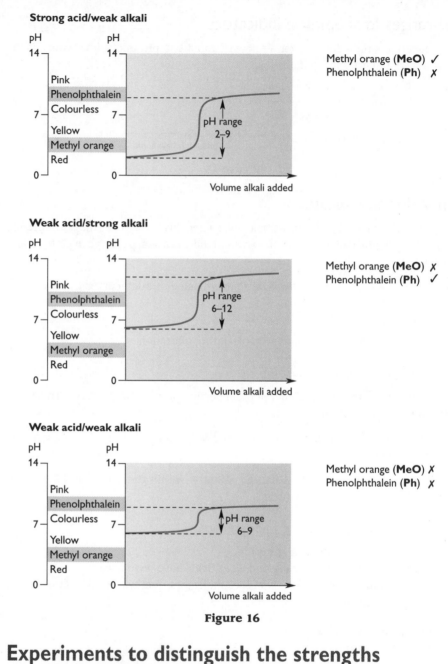

Figure 16

Experiments to distinguish the strengths of acids and bases

A measurement of the pH of a known concentration of an acid or a base would indicate whether it was strong or weak, but there are other ways in which this can be recognised. The simplest is to measure the ease with which they conduct electricity. In solution, ions conduct electricity and they are most effective if the ions are present in a high concentration. It follows that if solutions of strong and weak acids and bases are made with the same concentration in each case, the strong acids or bases will conduct better than those that are weak. The difference can therefore be easily identified.

Enthalpy of neutralisation can also be used to distinguish between strong and weak acids, and is defined in terms of formation of water.

The reason that it is defined in terms of the amount of water formed is to take account of the basicity of the acid. If the acid is dibasic, 1 mol of sulfuric acid would require twice as much sodium hydroxide to neutralise it compared to 1 mol of hydrochloric acid:

$$H_2SO_4(aq) + 2NaOH(aq) \rightarrow Na_2SO_4(aq) + 2H_2O(l)$$

$$HCl(aq) + NaOH(aq) \rightarrow NaCl(aq) + H_2O(l)$$

But if the enthalpy of neutralisation is quoted for the formation of 1 mol of water then this difference is removed and the figures obtained for the enthalpy change in the reactions can be directly compared. In fact, for the two acids above it turns out that the enthalpy change per mol of water produced is the same.

This is not really surprising because the reaction that is taking place is the same, i.e. $H^+(aq) + OH^-(aq) \rightarrow H_2O(l)$.

If the experiment is repeated using ethanoic acid and sodium hydroxide the enthalpy of neutralisation obtained is less. Since the reaction is the same the difference must lie in the strength of the acid. The dissociation of ethanoic acid is not complete:

$$CH_3COOH(aq) \rightleftharpoons CH_3COO^-(aq) + H^+(aq).$$

As the H^+ is neutralised by the base the equilibrium moves to the right and this requires a certain amount of energy such that some of the heat that would have been released is used up in this way.

The neutralisation of aqueous ammonia and hydrochloric acid gives an even lower value for the enthalpy of neutralisation because the process:

$$NH_3(g) + H_2O(l) \rightleftharpoons NH_4^+(aq) + OH^-(aq)$$

requires even more energy than the dissociation of the ethanoic acid.

Of course, the enthalpy of neutralisation of the weak acid/weak base pairing of ethanoic acid and ammonia would produce a further reduction in the value obtained.

Experiments of this type can give some indication of the strengths of acids and bases although such measurements are not particularly accurate.

The standard **enthalpy of neutralisation** is defined as the enthalpy change that occurs when 1 mol of water is produced in the reaction of an acid with an alkali under standard conditions.

Examiner tip

If asked to define the enthalpy of neutralisation of HCl(aq) it would be acceptable to state '*the enthalpy change when 1 mol of HCl(aq) is neutralised by an alkali under standard conditions*'. If asked the same question about $H_2SO_4(aq)$ then '*the enthalpy change when 1 mol of $H_2SO_4(aq)$ is neutralised by an alkali under standard conditions*' would be incorrect as this neutralisation would produce two moles of water.

Examiner tip

You might be asked to calculate the enthalpy of neutralisation and be expected to use:

$$\frac{Q}{n} = \frac{mc\Delta T}{n}$$

and you are strongly recommended to revise this section in Unit F322.

Having revised **Module 1: Rates, Equilibrium and pH** you should now have an understanding of:
- orders of reaction and rate equations
- rate-determining step
- equilibrium, K_c
- acids, bases and conjugate pairs
- pK_a and pH
- buffers
- enthalpy of neutralisation

Summary

Module 2: Energy

Lattice enthalpy

Review of basic ideas on energetics

Chemical reactions are usually accompanied by a change in enthalpy (energy), usually in the form of heat energy. Reactions tend to be either exothermic or endothermic.

If the reaction mixture loses energy to its surroundings, the reaction is exothermic and ΔH is negative.

If the reaction mixture gains energy from its surroundings, the reaction is endothermic and ΔH is positive.

Standard enthalpy changes

All standard enthalpy changes are measured under standard conditions. The temperature and the pressure at which measurements and/or calculations are carried out are standardised:

- standard temperature = 298 K (25°C)
- standard pressure = 100 kPa (100 000 N m^{-2} = 10^5 Pa = 1 bar = 1 atmosphere)
- standard temperature and pressure are often referred to as STP.

Examinations often ask for a definition of enthalpy changes and it is advisable to learn the following definitions.

- **Standard enthalpy change of formation** is the enthalpy change when 1 mol of a substance is formed from its elements, in their natural state, under standard conditions of 298 K and 100 kPa.
- **Standard enthalpy change of combustion** is the enthalpy change when 1 mol of a substance is burnt completely, in an excess of oxygen, under standard conditions of 298 K and 100 kPa.
- **Average bond enthalpy** is the enthalpy change on breaking 1 mol of a covalent bond in a gaseous molecule under standard conditions of 298 K and 100 kPa.

In addition to the definitions you may also be expected to show your understanding by writing equations to illustrate both the standard enthalpy change of formation and of combustion.

An equation to show the standard **enthalpy of formation** of ethane is given in Figure 17.

1 mole of product must always be formed even if it means using fractions in the balanced equation

$$2C(s) + 3H_2(g) \longrightarrow 1C_2H_6(g)$$

It is essential to show all state symbols

Figure 17

An equation to show the standard **enthalpy of combustion** of ethane is given in Figure 18.

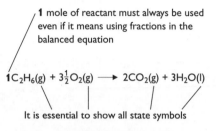

$$1C_2H_6(g) + 3\tfrac{1}{2}O_2(g) \longrightarrow 2CO_2(g) + 3H_2O(l)$$

1 mole of reactant must always be used even if it means using fractions in the balanced equation

It is essential to show all state symbols

Figure 18

Average bond (dissociation) enthalpy is the breaking of bonds and is always endothermic ($+\Delta H$) and is equivalent to homolytic fission. The breaking of the bond produces two neutral particles:

$$H–Cl(g) \rightarrow H(g) + Cl(g)$$

Bond enthalpies are the average (mean) values and take into account the chemical environment. In a molecule like water it is possible to successively break each of the two bonds and the energy involved will differ for each bond (Figure 19).

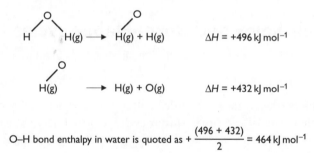

O–H bond enthalpy in water is quoted as $+\dfrac{(496 + 432)}{2} = 464\ kJ\,mol^{-1}$

Figure 19

Activation energy

Activation energy is defined as the minimum energy required, in a collision between particles, if they are to react. In any chemical reaction bonds have to be broken and new bonds have to be formed. Breaking bonds is an endothermic process requiring energy. This energy requirement contributes to the activation energy of a reaction.

Hess's law

Hess's law states that the enthalpy change for a reaction is the same irrespective of route taken provided that the initial and final conditions are the same.

Calculate the enthalpy change, ΔH_r, for the reaction $2CO(g) + O_2(g) \rightarrow 2CO_2(g)$. The standard enthalpies of formation of $CO(g)$ and $CO_2(g)$ are -110 and $-394\ kJ\,mol^{-1}$ respectively.

This is shown in Figure 20.

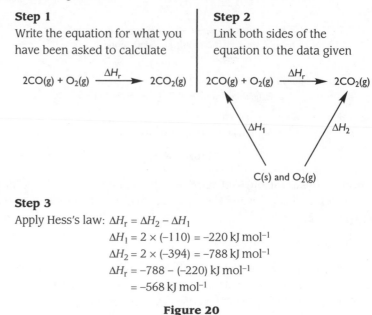

Step 1
Write the equation for what you have been asked to calculate

$$2CO(g) + O_2(g) \xrightarrow{\Delta H_r} 2CO_2(g)$$

Step 2
Link both sides of the equation to the data given

$$2CO(g) + O_2(g) \xrightarrow{\Delta H_r} 2CO_2(g)$$

ΔH_1 ΔH_2

$C(s)$ and $O_2(g)$

Step 3
Apply Hess's law: $\Delta H_r = \Delta H_2 - \Delta H_1$
$$\Delta H_1 = 2 \times (-110) = -220 \text{ kJ mol}^{-1}$$
$$\Delta H_2 = 2 \times (-394) = -788 \text{ kJ mol}^{-1}$$
$$\Delta H_r = -788 - (-220) \text{ kJ mol}^{-1}$$
$$= -568 \text{ kJ mol}^{-1}$$

Figure 20

Lattice enthalpy and Born–Haber cycles

The **lattice enthalpy** ($\Delta H_{LE}^{\ominus}$) of an ionic compound is the enthalpy change that accompanies the formation of 1 mol of an ionic compound from its constituent gaseous ions. ($\Delta H_{LE}^{\ominus}$ is exothermic.)

Lattice enthalpy indicates the strength of the ionic bonds in an ionic lattice.

$$Na^+(g) + Cl^-(g) \rightarrow Na^+Cl^-(s)$$

It is almost impossible to measure lattice enthalpy experimentally such that lattice enthalpy is calculated using a **Born–Haber** cycle. A Born–Haber cycle is similar to a Hess's cycle and enables the calculation of changes that cannot be measured directly by experiment.

The lattice enthalpy of sodium chloride can be calculated by considering the standard enthalpy of formation of NaCl(s). In order to form an ionic solid, both sodium and chlorine have to undergo a number of changes. These are outlined in Figure 21.

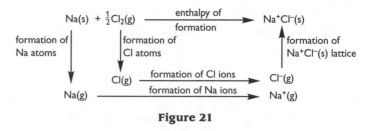

Figure 21

All of the changes in the cycle can be measured experimentally except the formation of the Na⁺Cl⁻(s) lattice from its gaseous ions, i.e. the lattice enthalpy. However, since all other steps in the cycle can be measured the lattice enthalpy can be calculated. This is done by adapting the above cycle and changing it into a Born–Haber cycle. The Born–Haber cycle is a combination of an enthalpy profile diagram and a Hess's cycle. The full Born–Haber cycle for sodium chloride is shown in Figure 22.

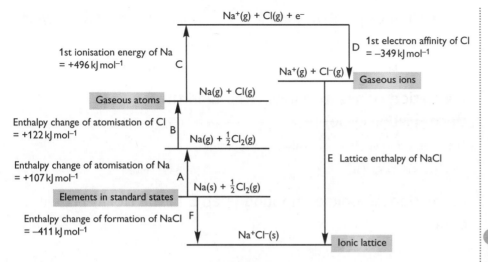

Figure 22

Using Hess's law:

A + B + C + D + E = F

$\Delta H_{at}^{\ominus}\,Na(g) + \Delta H_{at}^{\ominus}\,Cl(g) + \Delta H_{IE}^{\ominus}\,Na(g) + \Delta H_{EA}^{\ominus}\,Cl(g) + E = \Delta H_f^{\ominus}\,Na^+Cl^-(s)$

$\therefore\ +107 + 122 + 496 + (-349) + E = -411$

Hence, the lattice energy of $Na^+Cl^-(s)$, $E = -787\,kJ\,mol^{-1}$.

Definitions for enthalpy changes

Formation of an ionic compound (step F in the Born–Haber cycle)

The **standard enthalpy change of formation** is usually exothermic for an ionic compound:

$Na(s) + \frac{1}{2}Cl_2(g) \rightarrow Na^+Cl^-(s)$ $\Delta H_f^{\ominus} = -411\,kJ\,mol^{-1}$

Formation of gaseous atoms (steps A and B in the Born–Haber cycle)

The **standard enthalpy change of atomisation** is always endothermic:

$Na(s) \rightarrow Na(g)$ $\Delta H_{at}^{\ominus} = +107\,kJ\,mol^{-1}$

$\frac{1}{2}Cl_2(g) \rightarrow Cl(g)$ $\Delta H_{at}^{\ominus} = +122\,kJ\,mol^{-1}$

For gaseous molecules this enthalpy change can be determined from the **bond dissociation enthalpy** — the enthalpy change required to break and separate one mole of bonds so that the resulting gaseous atoms exert no forces upon each other:

$Cl-Cl(g) \rightarrow 2Cl(g)$ $\Delta H_{B.D.E.}^{\ominus} = +244\,kJ\,mol^{-1}$

$\frac{1}{2}Cl-Cl(g) \rightarrow Cl(g)$ $\Delta H_{at}^{\ominus} = +122\,kJ\,mol^{-1}$

Examiner tip

In most exams there is *at least* one question that asks for a definition; this can be worth as many as 3 marks. The difference between grade boundaries can sometimes be as low as 6 marks, so make sure you learn the definitions.

The **standard enthalpy change of formation**, $\Delta H_f^{\ominus}$ is the enthalpy change that takes place when 1 mol of a compound in its standard state is formed from its constituent elements in their standard states under standard conditions.

The **standard enthalpy change of atomisation**, $\Delta H_{at}^{\ominus}$, of an element is the enthalpy change that accompanies the formation of 1 mol of gaseous atoms from the element in its standard state.

The **first ionisation energy**, $\Delta H_{IE}^{\ominus}$, of an element is the enthalpy change that accompanies the removal of one electron from each atom in 1 mol of gaseous atoms to form 1 mol of gaseous I^+ ions.

Formation of positive ions (step C in the Born–Haber cycle)

The **first ionisation energy** is always endothermic:

$$Na(g) \rightarrow Na^+(g) + e^- \qquad\qquad \Delta H_{IE}^{\ominus} = +496\,kJ\,mol^{-1}$$

Formation of negative ions (step D in the Born–Haber cycle)

The **first electron affinity** is always exothermic:

$$Cl(g) + e^- \rightarrow Cl^-(g) \qquad\qquad \Delta H_{EA}^{\ominus} = -349\,kJ\,mol^{-1}$$

The **first electron affinity**, $\Delta H_{EA}^{\ominus}$, of an element is the enthalpy change that accompanies the addition of one electron to each atom in 1 mol of gaseous atoms to form 1 mol of gaseous I^- ions.

Formation of ionic compound (step E in the Born–Haber cycle)

$$Na^+(g) + Cl^-(g) \rightarrow Na^+Cl^-(s)$$

Examiner tip
When constructing a Born–Haber cycle it is essential to show the change from *elements* to *gaseous atoms* to *gaseous ions* to *ionic lattice* (Figure 23).

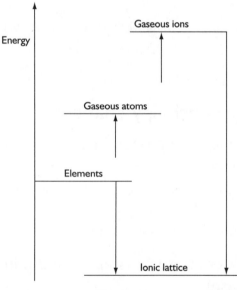

Figure 23

Calculation of lattice enthalpy

The **lattice enthalpy**, $\Delta H^{\ominus}_{LE}$, of an ionic compound is the enthalpy change that accompanies the formation of 1 mol of an ionic compound from its constituent gaseous ions ($\Delta H^{\ominus}_{LE}$ is exothermic).

The **lattice enthalpy** for magnesium chloride and for magnesium oxide can be calculated using the following data.

Stage	Standard enthalpy change	Equation	$\Delta H/kJ\,mol^{-1}$
A	Formation of $MgCl_2(s)$	$Mg(s) + Cl_2(g) \rightarrow MgCl_2(s)$	−641
B	Formation of $MgO(s)$	$Mg(s) + \tfrac{1}{2}O_2(g) \rightarrow MgO(s)$	−602
C	Atomisation of magnesium	$Mg(s) \rightarrow Mg(g)$	+148

Stage	Standard enthalpy change	Equation	$\Delta H/\text{kJ mol}^{-1}$
D	Atomisation of chlorine	$\frac{1}{2}Cl_2(g) \rightarrow Cl(g)$	+122
E	Atomisation of oxygen	$\frac{1}{2}O_2(g) \rightarrow O(g)$	+249
F	First ionisation energy of Mg	$Mg(g) \rightarrow Mg^+(g) + 1e^-$	+738
G	Second ionisation energy of Mg	$Mg^+(g) \rightarrow Mg^{2+}(g) + 1e^-$	+1451
H	First electron affinity of Cl	$Cl(g) + 1e^- \rightarrow Cl^-(g)$	−349
I	First electron affinity of O	$O(g) + 1e^- \rightarrow O^-(g)$	−141
J	Second electron affinity of O	$O^-(g) + 1e^- \rightarrow O^{2-}(g)$	+798

The Born–Haber cycle for magnesium chloride is shown in Figure 24.

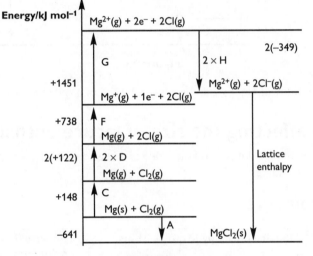

Figure 24

Applying Hess's law to the Born–Haber cycle gives:

A = C + 2D + F + G + 2H + LE

so, LE = A − C − 2D − F − G − 2H

The lattice enthalpy of magnesium chloride is:

= −641 − 148 − 244 − 738 − 1451 − (−689)

= −2524 kJ mol^{-1}

The Born–Haber cycle for magnesium oxide (Figure 25) is similar but in order to form the oxide ion, O^{2-}, the oxygen atom needs to gain two electrons and consequently has two electron affinities. The first electron affinity is exothermic but the second electron affinity is endothermic.

Examiner tip

When constructing a Born–Haber cycle for MgCl$_2$ students often lose marks by forgetting to double the value for the atomisation energy and the electron affinity of Cl.

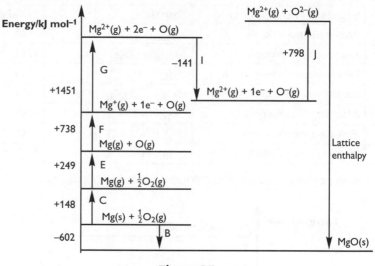

Figure 25

Knowledge check 19

Write equations to illustrate:

(a) atomisation energy of Cl_2

(b) lattice enthalpy of $CaCl_2$

(c) second ionisation energy of Mg

The lattice enthalpy of MgO(s) can be shown, by calculation, to be $3\,845\,kJ\,mol^{-1}$.

Factors affecting the size of lattice enthalpies

The strength of an ionic lattice and the value of its lattice enthalpy depend upon ionic radius and ionic charge.

Effect of ionic size

Compound	Lattice enthalpy/kJ mol^{-1}	Ions	Effect of ionic radius of halide ion
NaCl	−787		Ionic radius increases
NaBr	−751		Charge density decreases
NaI	−705		Attraction between ions decreases Lattice energy becomes less negative

Note that lattice energy has a negative value and you should use the term *'becomes less/more negative'* instead of *'becomes bigger/smaller'* to describe lattice enthalpies.

Effect of ionic charge

The strongest ionic lattices contain **small, highly charged ions**.

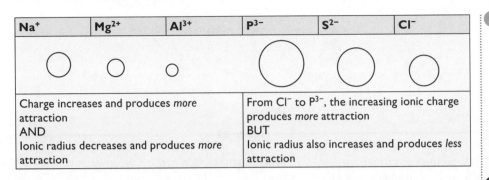

Na⁺	Mg²⁺	Al³⁺	P³⁻	S²⁻	Cl⁻

Charge increases and produces *more* attraction AND Ionic radius decreases and produces *more* attraction	From Cl⁻ to P³⁻, the increasing ionic charge produces *more* attraction BUT Ionic radius also increases and produces *less* attraction

Examiner tip
If asked to predict the size of lattice enthalpy it is best to avoid using terms like *bigger* or *smaller*. Lattice enthalpy is always negative and it is best to use terms such as *more negative* or *less negative*.

Knowledge check 20
Arrange the following in order of size of lattice enthalpy with the most negative first:
(a) NaBr, KBr, MgBr₂
(b) CaCl₂, BaCl₂, CaF₂, BaBr₂

The **enthalpy of hydration** of an ion is defined as the enthalpy change that occurs when 1 mol of gaseous ions are completely hydrated by water.

Enthalpy change of hydration

Solubility and enthalpies of hydration

The concept of a Born–Haber cycle can be extended to provide a partial explanation of the solubility of substances in water. To understand this another new enthalpy term needs to be introduced. This is the enthalpy of hydration of an ion which is defined as the enthalpy change that occurs when 1 mol of gaseous ions is completely hydrated by water.

It is therefore the enthalpy change for the process:

$$X^{n+}(g) \rightarrow X^{n+}(aq)$$

The standard **enthalpy of hydration** is quoted under the usual conditions of 25°C and 101 kPa.

In the case of hydration the attraction is either between a cation and the oxygen atom of a water molecule or between an anion and the hydrogen atom of the water molecule (Figure 26). This occurs because of the dipoles present in water, which you should be familiar with from the AS course.

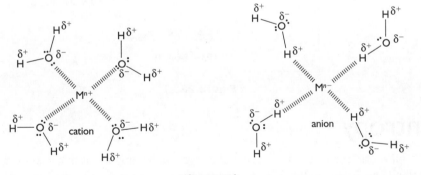

Figure 26

Like lattice enthalpy, the enthalpy change of hydration also depends on the ionic radius and the size of the charge of the ion and as with lattice enthalpy, the greater the charge density the greater the attraction.

Values of lattice enthalpies and enthalpies of hydration relate to the **enthalpy of solution**, which is the enthalpy change that occurs when an ionic solid dissolves in water. A typical enthalpy cycle for sodium chloride is shown in Figure 27.

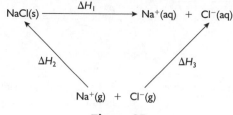

Figure 27

The **enthalpy of solution** of a compound is defined as the enthalpy change when I mol of that compound dissolves completely in excess water.

Applying Hess's law, $\Delta H_2 + \Delta H_1 = \Delta H_3$, such that $\Delta H_1 = \Delta H_3 - \Delta H_2$, where:

- ΔH_1 = enthalpy of solution
- ΔH_2 = lattice enthalpy of $NaCl(s)$ = $-781\,kJ\,mol^{-1}$
- ΔH_3 = enthalpy of hydration of the Na^+ ion (-418) + enthalpy of hydration of the Cl^- ion (-338) = $-756\,kJ\,mol^{-1}$

$$\Delta H_1 = \Delta H_3 - \Delta H_2 = -756 + 781 = +25\,kJ\,mol^{-1}$$

It might seem odd that the dissolving of sodium chloride is endothermic and yet sodium chloride readily dissolves in water at 25°C. It suggests that there is some other factor that is encouraging the dissolving to take place. It is an energy term called **entropy** and this is discussed in the next section.

> **Knowledge check 21**
>
> Write equations to illustrate the following:
>
> (a) the enthalpy of hydration of calcium ions
>
> (b) the enthalpy of solution of $Ca(OH)_2(s)$
>
> (c) the lattice enthalpy of $Ca(OH)_2(s)$

Examiner tip

This could be tested by providing you with a Born–Haber diagram and asking you to deduce the value of any one step. The Born–Haber diagram for NaCl is shown in Figure 28.

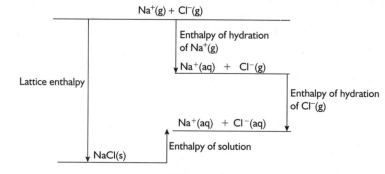

Figure 28

Entropy

Hess's law and the use of the Born–Haber cycle allow important information to be obtained concerning enthalpy changes that occur as a result of chemical reactions. But knowledge of enthalpy changes alone is insufficient to provide a certain answer as to whether a reaction will take place or not. Further information is needed about another energy change that takes place during a reaction. The information required is known as entropy.

When a reaction occurs enthalpy is either absorbed (endothermic) or released (exothermic) in the form of heat. In addition some energy is either absorbed or released as a result of the re-distribution of the particles when the products are formed. The extent of this energy depends largely on the physical state of the substances and on the temperature. Entropy is the term used to measure this quantity of energy and it is given the symbol S. We will be concerned in this course only with the reaction itself (known as the reaction system) although a more detailed study would include the effect on the surroundings.

Solids are more ordered than liquids, while gases are the least ordered and it follows that the most energy/entropy is required to hold a solid in its ordered state. The particles of a gas are less restrained than in a liquid and energy/entropy is not used up in restricting their freedom of movement.

The enthalpy change for the melting of ice to water at 0°C is:

$$H_2O(s) \rightarrow H_2O(l) \qquad \Delta H = +6.02 \, kJ \, mol^{-1}$$

The reaction therefore does not look possible, but we all know that ice does melt at 0°C. This can be explained because the change in entropy, as melting occurs, releases sufficient energy to counteract the positive enthalpy. The energy required to hold the rigid structure of the ice in place is released as the less restrained molecules of water are produced.

Enthalpy and entropy have a number of important differences:

Enthalpy	Entropy
If enthalpy is released when a reaction occurs ΔH is negative	If entropy is released when a reaction occurs ΔS is positive
The energy unit is usually in kilojoules, kJ	The energy unit is usually in joules, J

Entropy always increases (ΔS is positive) when there is a greater opportunity for energy to be spread out as a result of a change. It follows that entropy increases when any of the following occur.

- A solid becomes a liquid. Entropy is released and ΔS is positive — more disordered.
- A liquid becomes a gas. Entropy is released and ΔS is positive — more disordered.
- A solid dissolves in a liquid to form a solution. Entropy is released and ΔS is positive — more disordered.
- A reaction results in products with a greater degree of freedom of movement. For example, this could be because a gas is produced when a solid reacts, e.g. $CaCO_3(s) \rightarrow CaO(s) + CO_2(g)$. Entropy is released and ΔS is positive — more disordered.
- When a reaction produces more particles in the same state, e.g. $C_3H_8(g) + 5O_2(g) \rightarrow 3CO_2(g) + 4H_2O(g)$. Entropy is released and ΔS is positive — more disordered.
- The temperature rises, even if there is no change in state. Entropy is released and ΔS is positive — more disordered.

Calculating entropy changes

Calculations to determine ΔS are similar to calculations for ΔH, although it must be remembered that if ΔS is positive, it means the process releases entropy.

The change in entropy can be calculated using:

$$\Delta S = \Sigma(\text{entropy of products}) - \Sigma(\text{entropy of reactants})$$

Examiner tip

You will not be asked to explain what 'entropy' is. It is a difficult concept and even experts disagree. You may be asked to predict whether or not entropy increases in a reaction, in which case simply use the state symbols as a guide. Compare the number of moles of gas or liquid or solid on each side of the equation.

Knowledge check 22

For each of the following reactions predict whether the reaction will have a positive or negative value for the entropy change:

(a) $H_2O(g) \rightarrow H_2O(s)$

(b) $NaOH(s) \rightarrow NaOH(aq)$

(c) $2Mg(s) + O_2(g) \rightarrow 2MgO(s)$

(d) $2SO_2(g) + O_2(g) \rightarrow 2SO_3(g)$

Knowledge check 23

Calculate the entropy change when sodium reacts with oxygen.

$S^\ominus$ (sodium) = 51.0 J mol^{-1} K^{-1}; $S^\ominus$ (oxygen) = 102.5 J mol^{-1} K^{-1};

$S^\ominus$ (sodium oxide) = 72.8 J mol^{-1} K^{-1}

Calculate the entropy change for the reaction $3O_2(g) \rightarrow 2O_3(g)$ under standard conditions, given that $S^\ominus$ for O_3 is 238.8 J mol^{-1} K^{-1}, and $S^\ominus$ for O_2 is 205 J mol^{-1} K^{-1}.

$$\Delta S = \Sigma(\text{entropy of products}) - \Sigma(\text{entropy of reactants})$$

$$\Delta S = 2 \times (238.8) - 3 \times (205)$$

$$= -137.4 \text{ J mol}^{-1}\text{K}^{-1}$$

You should be able to anticipate the sign of ΔS by the fact that 3 mol of O_2 gas have changed to 2 mol of O_3 gas. When a reaction produces fewer particles in the same state, ΔS is negative.

Free energy

The change in the entropy of a reaction can be combined with the change in enthalpy to provide an answer to the question as to whether a chemical reaction is feasible.

A new term must be introduced called the free energy (strictly this is known as the Gibbs free energy) which is given the symbol G. The free energy change of a reaction relates to the enthalpy and entropy changes by the equation:

$$\Delta G = \Delta H - T\Delta S$$

and ΔG provides a certain answer as to whether a given reaction will be feasible. If ΔG is negative, the reaction will definitely be feasible and if ΔG is positive then, at least at the particular temperature chosen, the reaction will not be feasible.

Any reaction will fit one of four possible scenarios:
- **ΔH is negative and ΔS is positive**. The reaction will always be feasible.
- **ΔH is positive and ΔS is negative**. The reaction will never be feasible.
- **ΔH is negative and ΔS is negative**. ΔH favours the reaction, but $T\Delta S$ resists the change. The reaction will be feasible when $\Delta H > T\Delta S$ and is therefore more likely to be feasible at low temperatures.
- **ΔH is positive and ΔS is positive**. ΔH resists the reaction but $T\Delta S$ favours the change. The reaction will be feasible when $\Delta H < T\Delta S$ and is therefore more likely to be feasible at high temperatures.

Equilibrium

If $\Delta G = 0$, then the system will be at equilibrium and $\Delta H = T\Delta S$. The enthalpy change for the melting of ice to water at 0°C is $H_2O(s) \rightarrow H_2O(l)$ $\Delta H = +6.02$ kJ mol^{-1}. At 0°C (273 K) $\Delta G = 0$ such that $\Delta S = \Delta H/T$, hence the entropy change when ice melts is:

$$\Delta S = \frac{\Delta H}{T} = \frac{6.02 \text{ kJ mol}^{-1}}{273 \text{ K}}$$

but entropy is measured in J mol^{-1} K^{-1}, so:

$$\frac{6020 \text{ J mol}^{-1}}{273 \text{ K}} = 22.0 \text{ J mol}^{-1}\text{K}^{-1}$$

For a chemical reaction, the values of ΔH and ΔS must be calculated and then the value of the temperature, T, can be established for which ΔG is zero. It should first be noted, however, that equilibrium can never be reached for a reaction for which either

ΔH is negative and ΔS is positive (because the reaction is always feasible) or ΔH is positive and ΔS is negative (because the reaction is never feasible), but where ΔH and ΔS have the same sign it will be possible to find the equilibrium temperature noting that as $\Delta G = 0$, $\Delta H = T\Delta S$, so:

$$T = \frac{\Delta H}{\Delta S}$$

Examiner tip
ΔG is only temperature dependent when ΔH and ΔS have the same sign.

Calculating free energy changes

If tables of information are provided, then calculating the value of ΔG for a reaction is identical to the process of calculating ΔH.

If values for ΔH and S are available for each component in the equation then it is a little more laborious but ΔG can be calculated for each substance in turn and then the overall ΔG for the reaction can then be determined.

Example

Use the data below to calculate the temperature at which the reaction $2NO(g) + O_2(g) \rightarrow 2NO_2(g)$ reaches equilibrium.

	$\Delta H_f^{\ominus}$/kJ mol^{-1}	$S^{\ominus}$/J mol^{-1} K^{-1}
$NO(g)$	90.4	210.5
$O_2(g)$	0	204.9
$NO_2(g)$	33.2	240.0

Use:
- $\Delta H = \Sigma$(enthalpy of products) $- \Sigma$(enthalpy of reactants)
- $\Delta S = \Sigma$(entropy of products) $- \Sigma$(entropy of reactants)
- at equilibrium, $\Delta G = 0$, so $T = \Delta H/\Delta S$

ΔH for the reaction is:

$$(2 \times 33.2) - (2 \times 90.4) = -114.4 \text{ kJ mol}^{-1}$$

and

$$\Delta S = (2 \times 240.0) - [(2 \times 210.5) + 204.9] = -145.9 \text{ J mol}^{-1}\text{K}^{-1}$$

Therefore (remembering to convert ΔS from J into kJ),

$$T = -114.4/-0.1459 = 784 \text{ K or } 511°C$$

Examiner tip
When using $\Delta G = \Delta H - T\Delta S$ you must remember that the units of ΔG and ΔH are both kJ mol^{-1} *but* ΔS is measured in J mol^{-1} K^{-1}. Therefore ΔS has to be converted into kJ mol^{-1} K^{-1}.

Electrode potentials and fuel cells

Ionic equations

To write ionic equations correctly, it is essential to remember the following points.
- Ionic substances that are solid do not have free-moving ions and therefore their ions cannot react independently of each other. In an ionic equation their complete formulae must be given.

- Compounds of metals and strong acids in aqueous solution will always be split into their ions. These ions can, and do, react independently of each other.
- Covalent compounds exist as complete molecules and are always shown as complete entities in the ionic equation.
- It is important to include the state symbols in all ionic equations.
- It is essential to balance both symbols and charge.

Acid + base → salt + water

When an aqueous hydroxide reacts with an acid, the ionic equation summarises this as:

$$H^+(aq) + OH^-(aq) \rightarrow H_2O(l)$$

Both sides of the equation have a net charge of zero.

When a base like magnesium oxide reacts with an acid the ionic equation becomes:

$$MgO(s) + 2H^+(aq) \rightarrow Mg^{2+}(aq) + H_2O(l)$$

Both sides of the equation have a net charge of 2+.

Acid + base → salt + carbon dioxide + water

When an aqueous carbonate reacts with an acid, the ionic equation is written as:

$$CO_3^{2-}(aq) + 2H^+(aq) \rightarrow CO_2(g) + H_2O(l)$$

Both sides of the equation have a net charge of zero.

Redox reactions and oxidation numbers

The displacement reaction between chlorine and bromide (Figure 29) is an example of a redox reaction that you met in Unit F321 at AS:

$$Cl_2(g) + 2Br^- (aq) \rightarrow 2Cl^-(aq) + Br_2(aq)$$

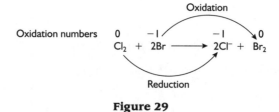

Figure 29

Remember that **o**xidation **i**s **l**oss, **r**eduction **i**s **g**ain ('OILRIG')

Oxidation number

Oxidation number is a convenient way of quickly identifying whether or not a substance has undergone either oxidation or reduction. In order to work out the oxidation number you must first learn a few simple rules.

Rule	Example
I	All elements in their natural state have oxidation number = 0 H_2, oxidation number of H = 0
2	The oxidation numbers of any molecule always add up to zero H_2O, sum of oxidation numbers = 0
3	The oxidation numbers of any ion always add up to the charge of the ion SO_4^{2-}, sum of oxidation numbers = −2
When calculating the oxidation numbers of elements in either a molecule or an ion you should apply the following order of priority.	
(i)	Groups I, 2 and 3 elements are always +1, +2 and +3, respectively
(ii)	Fluorine is always −1
(iii)	Hydrogen is usually +1
(iv)	Oxygen is usually −2
(v)	Chlorine is usually −1

If you make sure you apply these rules rigidly in the sequence indicated it should be relatively simple to deduce any oxidation number.

Consider the reaction of zinc and aqueous copper sulfate:

$$Zn(s) + CuSO_4(aq) \rightarrow ZnSO_4(aq) + Cu(s)$$

It is often helpful to immediately write the oxidation numbers above each element in the equation:

Oxidation numbers

0		+2 +6 −2		+2 +6 −2		0
Zn(s)	+	$CuSO_4(aq)$	$\rightarrow$	$ZnSO_4(aq)$	+	Cu(s)

In any redox reaction the oxidation number of one element increases while the oxidation number of a second element decreases. Zn increases from 0 to +2, while Cu decreases from +2 to 0 such that the ionic equation can be written as:

$$Zn(s) + Cu^{2+}(aq) \rightarrow Zn^{2+}(aq) + Cu(s)$$

The Zn is oxidised, losing two electrons and becoming a Zn^{2+} ion. The electrons are taken up by the Cu^{2+} ion as it is reduced to Cu metal. It is possible to write ionic half-equations:

$$Zn(s) \rightarrow Zn^{2+}(aq) + 2e^- \qquad \text{Oxidation (loss of electrons)}$$

$$Cu^{2+}(aq) + 2e^- \rightarrow Cu(s) \qquad \text{Reduction (gain of electrons)}$$

It should also be possible to use ionic half-equations to construct a full ionic equation. When Cu(s) is added to aqueous $AgNO_3(aq)$, $Cu^{2+}(aq)$ and Ag(s) are formed such that the ionic half-equations are:

$$Cu(s) \rightarrow Cu^{2+}(aq) + 2e^- \qquad \text{which is oxidation (loss of electrons), and}$$

$$Ag^+(aq) + e^- \rightarrow Ag(s) \qquad \text{which is reduction (gain of electrons).}$$

Examiner tip

When you have balanced an equation *always* double-check to make sure that the total charge on each side of the equation balances as well as the symbols.

In any two ionic half-equations the number of electrons released (by oxidation) must be the same as the number required for the reduction. Cu(s) supplies two electrons as it oxidises to Cu^{2+}(aq) but each Ag^+(aq) requires only one electron to be reduced to Ag(s). Therefore it will be necessary to use two Ag^+(aq) for every one Cu(s) such that:

$$Cu(s) \rightarrow Cu^{2+}(aq) + 2e^-$$

$$2Ag^+(aq) + 2e^- \rightarrow 2Ag(s)$$

The overall equation is then obtained by adding the two half-equations together but excluding the electrons:

$$2Ag^+(aq) + Cu(s) \rightarrow 2Ag(s) + Cu^{2+}(aq)$$

Knowledge check 24

Use each of the following pairs of half-equations to construct an overall equation for the reaction. You *must* balance each half-equation before constructing the overall equation.

(a) MnO_4^-(aq) + H^+(aq) → Mn^{2+}(aq) + H_2O(l) V^{2+}(aq) → V^{3+}(aq)

(b) MnO_4^-(aq) + H^+(aq) → Mn^{2+}(aq) + H_2O(l) V^{2+}(aq) + H_2O → VO_3^-(aq) + H^+(aq)

(c) $Cr_2O_7^{2-}$(aq) + H^+(aq) → Cr^{3+}(aq) + H_2O(l) SO_2(aq) + H_2O(l) → SO_4^{2-}(aq) + H^+(aq)

(d) NO_3^-(aq) → NO(g)   Cu(s) → Cu^{2+}(aq)

Electrode potentials

The **standard electrode potential** is the potential difference (the difference in voltage) between one half-cell (e.g. a metal in contact with its metal ions) and the standard hydrogen electrode when measured under standard conditions ($T = 298\,K$, $P = 101\,kPa$ and concentration $= 1.0\,mol\,dm^{-3}$).

The **standard cell potential** is the voltage formed when two half-cells are connected and the voltage measured using a voltmeter of very high resistance, under standard conditions ($T = 298\,K$, $P = 101\,kPa$ and concentration $= 1.0\,mol\,dm^{-3}$).

The **standard electrode potentials** of two half-cells can be combined to give a **standard cell potential**. This is important and it defines whether or not a reaction is feasible.

The standard hydrogen electrode is shown in Figure 30.

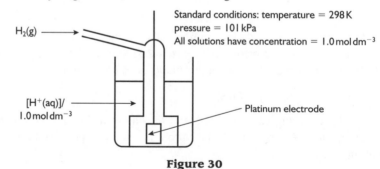

Figure 30

To provide a better surface for the hydrogen, the platinum electrode is usually coated with very finely divided platinum known as platinum black. This cell is connected via an external circuit and through a salt bridge to the other cell. The voltage measured then gives what is known as the electrode potential of the cell on a scale with the half-reaction $2H^+$(aq) + $2e^- \rightarrow H_2$(g) being given the arbitrary value of zero.

Measure the standard electrode potentials

For metals the system shown in Figure 31 applies.

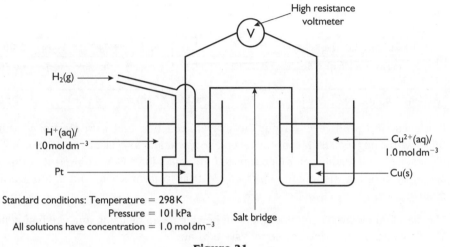

Standard conditions: Temperature = 298 K
Pressure = 101 kPa
All solutions have concentration = 1.0 mol dm^{-3}

Figure 31

For non-metals/ions of the same element in different oxidation states the system in Figure 32 applies:

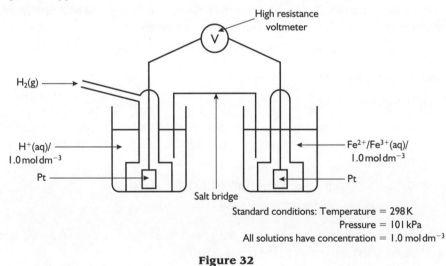

Standard conditions: Temperature = 298 K
Pressure = 101 kPa
All solutions have concentration = 1.0 mol dm^{-3}

Figure 32

The salt bridge is made of a porous material soaked in a saturated solution of KNO_3. The salt bridge completes the circuit without mixing the solutions by allowing the passage of ions.

Some common cell potentials are listed in the table below.

Half-cell	$E^\ominus/V$	Half-cell	$E^\ominus/V$
$F_2(g) + 2e^- \rightleftharpoons 2F^-(aq)$	+2.87	$Fe^{3+}(aq) + e^- \rightleftharpoons Fe^{2+}(aq)$	+0.77
$MnO_4^-(aq) + 8H^+(aq) + 5e^- \rightleftharpoons Mn^{2+}(aq) + 4H_2O(l)$	+1.52	$Cu^{2+}(aq) + 2e^- \rightleftharpoons Cu(s)$	+0.34
$Cl_2(g) + 2e^- \rightleftharpoons 2Cl^-(aq)$	+1.36	$2H^+(aq) + 2e^- \rightleftharpoons H_2(g)$	0.00
$Cr_2O_7^{2-}(aq) + 14H^+(aq) + 6e^- \rightleftharpoons 2Cr^{3+}(aq) + 7H_2O(l)$	+1.33	$Zn^{2+}(aq) + 2e^- \rightleftharpoons Zn(s)$	-0.76
$Ag^+(aq) + e^- \rightleftharpoons Ag(s)$	+0.80	$K^+(aq) + e^- \rightleftharpoons K(s)$	-2.92

Half-cells with positive $E^\ominus$ favour the forward reaction and gain electrons, while those with a negative $E^\ominus$ favour the reverse reaction and lose electrons. $F_2(g)$ has the highest $+E^\ominus$ and readily gains electrons to form $F^-(aq)$ ions. F_2 is a powerful oxidising agent. By contrast $K(s)$ readily loses an electron to form $K^+(aq)$ and is a powerful reducing agent.

The standard electrode cell potential can be calculated by using any two half-cells. The $E^\ominus$ values for the zinc and copper systems are:

$$Zn^{2+}(aq) + 2e^- \rightleftharpoons Zn(s) \qquad\qquad -0.76\,V$$

$$Cu^{2+}(aq) + 2e^- \rightleftharpoons Cu(s) \qquad\qquad +0.34\,V$$

which indicates that the $Cu^{2+}(aq)$ favours the forward reaction while the $Zn(s)$ favours the reverse reaction. Each half-cell can now be re-written as:

$$Cu^{2+}(aq) + 2e^- \rightarrow Cu(s) \qquad\qquad +0.34\,V$$

$$Zn(s) \rightarrow Zn^{2+}(aq) + 2e^- \qquad\qquad +0.76\,V$$

$$\text{Cell potential} \qquad\qquad +1.10\,V$$

Acidified $H^+(aq)/MnO_4^-(aq)$ is a good oxidising agent and can be used to prepare $Cl_2(g)$ by the oxidation of $Cl^-(aq)$ ions. Calculate the cell potential and deduce the balanced equation.

$$MnO_4^-(aq) + 8H^+(aq) + 5e^- \rightarrow Mn^{2+}(aq) + 4H_2O(l) \qquad +1.52\,V$$

$$Cl_2(g) + 2e^- \rightarrow 2Cl^-(aq) \qquad +1.36\,V$$

Both electrode potentials are positive but $H^+(aq)/MnO_4^-(aq)$ is most positive and therefore more likely to move to the right such that:

$$MnO_4^-(aq) + 8H^+(aq) + 5e^- \rightarrow Mn^{2+}(aq) + 4H_2O(l) \qquad +1.52\,V$$

$$2Cl^-(aq) \rightarrow Cl_2(g) + 2e^- \qquad -1.36\,V$$

$$\text{Cell potential} \qquad +0.16\,V$$

Both half-equations must have the same number of electrons. MnO_4^- is multiplied by 2 to give $10e^-$ while Cl^- is multiplied by 5.

$$2MnO_4^-(aq) + 16H^+(aq) + 10e^- \rightarrow 2Mn^{2+}(aq) + 8H_2O(l)$$

$$10Cl^- \rightarrow 5Cl_2(g) + 10e^-$$

$$2MnO_4^-(aq) + 16H^+(aq) + 10Cl^- \rightarrow 2Mn^{2+}(aq) + 8H_2O(l) + 5Cl_2(g)$$

Knowledge check 25

Use the following standard electrode potentials:

	$E^{\ominus}$		$E^{\ominus}$
$Mg^{2+}(aq) + 2e^- \rightleftharpoons Mg(s)$	$-2.37\,V$	$Zn^{2+}(aq) + 2e^- \rightleftharpoons Zn(s)$	$-0.76\,V$
$Sn^{4+}(aq) + 2e^- \rightleftharpoons Sn^{2+}(aq)$	$+0.15\,V$	$I_2(aq) + 2e^- \rightleftharpoons 2I^-(aq)$	$+0.54\,V$
$Fe^{3+}(aq) + e^- \rightleftharpoons Fe^{2+}(aq)$	$+0.77\,V$	$Br_2(aq) + 2e^- \rightleftharpoons 2Br^-(aq)$	$+1.09\,V$

to calculate the cell potential for each of the following pairs of half-cells:

(a) $Mg^{2+}(aq)/Mg(s)$ and $Zn^{2+}(aq)/Zn(s)$

(b) $Sn^{4+}(aq)/Sn^{2+}(aq)$ and $Fe^{3+}(aq)/Fe^{2+}(aq)$

(c) $I_2(aq)/2I^-(aq)$ and $Br_2(aq)/2Br^-(aq)$

(d) $Zn^{2+}(aq)/Zn(s)$ and $I_2(aq)/2I^-(aq)$

(e) $Sn^{4+}(aq)/Sn^{2+}(aq)$ and $Br_2(aq)/2Br^-(aq)$

Effect of concentration on the feasibility of reactions

A positive cell potential indicates that a reaction is feasible but it gives no indication as to how fast a reaction will occur. The cell potential of the reaction between $H^+(aq)/MnO_4^-(aq)$ and $Cl^-(aq)$ is only $+0.16\,V$ but the reaction takes place quickly despite the low overall potential.

Cell potentials are calculated assuming standard conditions and if the concentration of one component in a half-cell is changed, according to le Chatelier's principle the equilibrium will shift in such a way as to minimise the effect of the change.

If, for example, the equilibrium $Fe^{3+}(aq) + e^- \rightleftharpoons Fe^{2+}(aq)$, for which $E^{\ominus} = +0.77\,V$, is carried out with a reduced concentration of $Fe^{2+}(aq)$ then this will encourage a movement from left to right in the equilibrium and this will then cause the value of the electrode potential to increase. If there is a reduced concentration of $Fe^{3+}(aq)$ then the equilibrium will move to the left and the value of $E^{\ominus}$ will decrease. In both cases, a very large change would be required to make any noticeable difference. As a general rule of thumb, a ten-fold change in concentration only changes the electrode potential of a half-reaction by $0.06\,V$ or less.

Storage cells and fuel cells

Storage cells

Storage cells are commonly referred to as batteries and are used in appliances to supply electricity. Electrode potential can be used to predict the possible voltage of a battery but there is no need to remember the details or constructions of any particular cell. You will not be asked in any exam to recall any particular type of cell but you may be expected to interpret data that has been provided.

Commercial batteries tend to avoid the use of solutions that can easily leak and commercial cells usually contain pastes that surround the electrodes. An alkaline battery has a cathode made from graphite and manganese(IV) oxide and an anode made either of zinc or nickel-plated steel. The half-cell that provides the electrons to the external circuit is called the anode and the half-cell that receives them is the cathode. The pastes or solutions are called the electrolyte. The electrolyte is potassium hydroxide.

The reactions that take place are:

- at the anode

  $Zn + 2OH^- \rightarrow ZnO + H_2O + 2e^-$

- at the cathode

  $2MnO_2 + H_2O + 2e^- \rightarrow Mn_2O_3 + 2OH^-$

The overall equation taking place can be found in the usual way by combining the two half-reactions:

  $Zn + 2MnO_2 \rightarrow ZnO + Mn_2O_3$

Both the OH^- and the H_2O can be eliminated from the equation and their concentrations should remain constant. The overall voltage is about 1.5 V and is a combination of the electrode potentials of the two half-reactions.

Another commonly encountered battery is the rechargeable nickel–cadmium (Ni–Cd) cell. While it is supplying electricity the reactions taking place are:

- at the anode

  $Cd + 2OH^- \rightarrow Cd(OH)_2 + 2e^-$

- at the cathode

  $2NiO(OH) + 2H_2O + 2e^- \rightarrow 2Ni(OH)_2 + 2OH^-$

Once again, the electrolyte is potassium hydroxide. The overall reaction is:

  $Cd + 2NiO(OH) + 2H_2O \rightarrow Cd(OH)_2 + 2Ni(OH)_2$

The battery can be recharged by applying an external voltage which reverses the reactions above. A disadvantage of this type of battery is that cadmium is toxic and care needs to be taken when disposing of the batteries. A voltage of around 1.2 V can be obtained during use.

Many other storage cells are made and the intense research into the development of batteries for use in vehicles has led to many different constructions. Currently the lead–acid accumulator is still widely employed but the disadvantage of its considerable weight is obvious. Alternatives such as a sodium/sulfur cell are much lighter but can be expensive and much more risky in use.

Fuel cells

Fuel cells can produce electrical power from the chemical reaction of a fuel (such as hydrogen, hydrocarbons or alcohols) with oxygen. The fuel cell operates like a conventional storage cell except that the fuels are supplied as gases externally. The cell will therefore operate more or less indefinitely so long as the supply is maintained.

The hydrogen/oxygen fuel cell (Figure 33) is used widely and illustrates the principles behind a fuel cell.

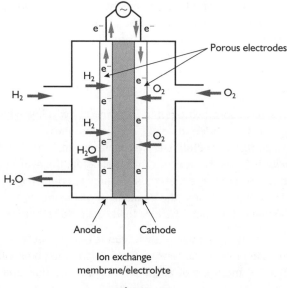

Figure 33

The electrodes are made of materials such as a titanium sponge coated in platinum. The electrolyte is an acid or alkaline membrane which will allow ions to move from one compartment of the cell to the other such that its acts as a salt bridge.

In an acidic solution, hydrogen is converted to hydrogen ions at the cathode while, at the anode, oxygen reacts with hydrogen ions to make water:

$$H_2(g) \rightleftharpoons 2H^+(aq) + 2e^- \qquad\qquad E^\ominus = 0\,V$$

$$\tfrac{1}{2}O_2(g) + 2H^+(aq) + 2e^- \rightleftharpoons H_2O(l) \qquad E^\ominus = +1.23\,V$$

The overall reaction is therefore $H_2(g) + \tfrac{1}{2}O_2(g) \rightarrow H_2O(l)$ and the voltage produced is 1.23 V.

In an alkaline solution, hydrogen reacts with hydroxide ions to form water while oxygen reacts with water to form hydroxide ions.

$$H_2(g) + 2OH^-(aq) \rightleftharpoons 2H_2O(l) + 2e^- \qquad E^\ominus = +0.83\,V$$

$$\tfrac{1}{2}O_2(g) + H_2O(l) + 2e^- \rightleftharpoons 2OH^-(aq) \qquad E^\ominus = +0.40\,V$$

The overall reaction is, as before, $H_2(g) + \tfrac{1}{2}O_2(g) \rightarrow H_2O(l)$ and, again, the voltage produced is 1.23 V.

Hydrogen–oxygen fuel cells offer an alternative to the use of fossil fuels (petrol or diesel), which will ultimately be depleted, and avoid the production of polluting products such as carbon monoxide, carbon dioxide or oxides of nitrogen. They are lightweight and offer a greater efficiency than engines based on fossil fuels.

There are drawbacks and problems that have to be overcome. Hydrogen is a gas and it is dangerously explosive. Possible solutions include compressing the gas until it liquefies, but the transportation of hydrogen under pressure is potentially hazardous. An alternative is to adsorb it on to the surface of a suitable solid material or to absorb it into a suitable material. A number of transition metal alloys have been tried.

Examiner tip

In all fuel cells the fuel is supplied to one electrode and $O_2(g)$ is supplied to the other electrode. The equation for the reaction at the O_2 electrode is either: $\tfrac{1}{2}O_2(g) + 2H^+(aq) + 2e^- \rightleftharpoons H_2O(l)$ or $\tfrac{1}{2}O_2(g) + H_2O(l) + 2e^- \rightleftharpoons 2OH^-(l)$ depending on whether the electrolyte is acidic or alkaline.

For example, an alloy of iron and titanium has been used but, although this absorbs the hydrogen quite well, it is too heavy to be entirely practical. Another possibility is to use carbon nanotubes. These are tiny lightweight structures containing arrangements of carbon atoms and which have small, cylindrical pores that capture the hydrogen. This material is certainly light enough but needs further investigation before it could be considered a commercial proposition.

Many metals can also absorb hydrogen to form metal hydrides, which can be made to release their hydrogen under the right conditions. Particular interest has been shown in the use of lighter metals such as magnesium, which forms the hydride MgH_2. A problem with this is that it needs to be hot before the hydrogen is released. In addition the manufacture of these materials requires energy which must be included in assessing their viability as must the expense of disposing of and replacing them once they have been used. In fact, hydrides tend to have a rather limited lifetime.

The major issue concerning the use of hydrogen is one of safety, although this also applies now to the use of hydrocarbons as fuels. There must be continuing research into safe and effective methods of production, transportation and storage of any hazardous materials. These issues along with finding methods of production of hydrogen that are economically and environmentally feasible will shape the possible use of hydrogen as a fuel for the future.

> **Examiner tip**
>
> Fuels other than hydrogen can also be used in fuel cells. In any fuel cell the electrons always flow (via the external circuit) from the electrode to which the fuel is supplied.

> **Knowledge check 26**
>
> Methanol, CH_3OH, can be used as the fuel in a fuel cell. Construct equations for the overall reaction and for the reactions at each electrode. Assume that an acidic electrolyte is used.

Summary

Having revised **Module 2: Energy** you should now have an understanding of:
- lattice enthalpy and Born–Haber cycles
- enthalpies of solution and hydration
- entropy, and the relationship between free energy, enthalpy and entropy
- electrode potentials
- storage cells and fuel cells

> A **transition element** is a d-block element that forms one or more stable ions that have partly filled d-orbitals.

Module 3: Transition elements

Properties of transition elements

The fourth period runs from K to Kr:

s-block		d-block										p-block					
K	Ca	Sc	Ti	V	Cr	Mn	Fe	Co	Ni	Cu	Zn	Ga	Ge	As	Se	Br	Kr

Electron configurations of d-block elements

The 4s-subshell is at a lower energy level than the 3d-subshell and therefore the 4s-subshell fills before the 3d-subshell. The orbitals in the 3d-subshell are first occupied singly to prevent any repulsion caused by pairing.

Filling the 4s- and 3d-subshells

		4s	3d					
Sc	[Ar] $3d^1$ $4s^2$	↑↓	↑					
Ti	[Ar] $3d^2$ $4s^2$	↑↓	↑	↑				
V	[Ar] $3d^3$ $4s^2$	↑↓	↑	↑	↑			
Cr	[Ar] $3d^5$ $4s^1$ *	↑	↑	↑	↑	↑	↑	
Mn	[Ar] $3d^5$ $4s^2$	↑↓	↑	↑	↑	↑	↑	
Fe	[Ar] $3d^6$ $4s^2$	↑↓	↑↓	↑	↑	↑	↑	
Co	[Ar] $3d^7$ $4s^2$	↑↓	↑↓	↑↓	↑	↑	↑	
Ni	[Ar] $3d^8$ $4s^2$	↑↓	↑↓	↑↓	↑↓	↑	↑	
Cu	[Ar] $3d^{10}$ $4s^{1}$**	↑	↑↓	↑↓	↑↓	↑↓	↑↓	
Zn	[Ar] $3d^{10}$ $4s^2$	↑↓	↑↓	↑↓	↑↓	↑↓	↑↓	

*Chromium has one electron in each orbital of the 4s- and 3d-subshells, giving the configuration [Ar] $3d^5$ $4s^1$, which is more stable than [Ar] $3d^4$ $4s^2$.

**Copper has a full 3d-subshell, giving the configuration [Ar] $3d^{10}$ $4s^1$, which is more stable than [Ar] $3d^9$ $4s^2$.

The majority of transition elements form ions in more than one oxidation state. When transition elements form ions they do so by losing electrons from the 4s-orbitals before the 3d-orbitals. Sc and Zn each form ions in one oxidation state only, Sc^{3+} and Zn^{2+}. The electron configurations of these ions are shown as [Ar] $3d^0$ and [Ar] $3d^{10}$ respectively, such that neither fits the definition of a transition element.

Typical properties of transition elements

The transition elements are all metals and, therefore, they are good conductors of heat and electricity.

Variable oxidation states

Transition elements have compounds with two or more oxidation states. This is primarily due to the fact that successive ionisation energies of transition metals only increase gradually. All the transition metals can form an ion of oxidation state +2, representing the loss of the two 4s-electrons. The maximum oxidation state possible cannot exceed the total number of 4s- and 3d-electrons in their electron configuration. (Once these electrons have been lost the stable electron configuration of argon is reached.)

Colour of compounds

The transition elements have at least one oxidation state in which their compounds and ions are coloured. The colours are often distinctive and can be used as a means of identification. $Cu^{2+}(aq)$ ions are blue, $Cr^{3+}(aq)$ ions are green, $Cr_2O_7^{2-}(aq)$ are orange while $MnO_4^-(aq)$ ions are purple.

Catalysis

Transition metals are frequently used as heterogeneous catalysts. This is a result of the use of their d-orbitals to bind other molecules or ions to their surface. Examples

Examiner tip

Most students get the full electron configuration of a transition metal correct. However, when asked for the electron configuration of a transition metal *ion* many incorrectly remove the 3d electrons before the 4s. The electron configuration of $_{26}Fe^{2+}$ is $1s^2$ $2s^2$ $2p^6$ $3s^2$ $3p^6$ $3d^6$, *not* $1s^2$ $2s^2$ $2p^6$ $3s^2$ $3p^6$ $3d^4$ $4s^2$.

Knowledge check 27

Deduce the oxidation number of the transition element in each of the following:

(a) $[Zn(NH_3)_4(H_2O)_2]^{2+}$

(b) $[Fe(CN)_6]^{4-}$

(c) $[Co(NH_3)_5Cl]^{2+}$

(d) $[Co(C_2O_4)_3]^{4-}$

(e) $[Cr(CH_3COO)_2(H_2O)_2]^+$

include the use of iron as a catalyst in the Haber process to produce ammonia, the use of platinum, palladium and rhodium in the catalytic converter of a car and the use of nickel in the hydrogenation of alkenes.

Complex ions with ligands

The ions of transition elements form complex ions with ligands.

Other properties

Other properties of transition elements are that they are denser than other metals. They have smaller atoms than the metals in groups 1 and 2 such that the atoms are able to pack closely together, hence increasing the density. They also have higher melting and boiling points than other metals. This can also be explained by considering the size of the atoms of the transition elements. Within the metallic lattice, ions are smaller than those of the s-block metals, which results in greater 'free electron density' and hence a stronger metallic bond.

Simple precipitation reactions

A precipitation reaction takes place between aqueous alkali and an aqueous solution of a metal(II) or metal(III) cation. This results in formation of a precipitate of the metal hydroxide, often with a characteristic colour. A suitable aqueous alkali is NaOH(aq). The colour of the precipitate can be used as a means of identification.

These precipitation reactions can be represented simply as follows:

$Cu^{2+}(aq) + 2OH^-(aq) \rightarrow Cu(OH)_2(s)$ Pale-blue precipitate

$Fe^{2+}(aq) + 2OH^-(aq) \rightarrow Fe(OH)_2(s)$ Pale-green gelatinous precipitate

$Fe^{3+}(aq) + 3OH^-(aq) \rightarrow Fe(OH)_3(s)$ Orange–brown gelatinous precipitate

$Fe(OH)_2(s)$ is slowly oxidised to $Fe(OH)_3(s)$ and, if left, the pale-green precipitate will change to orange–brown.

Transition element complexes

Ligands and complex ions

A **ligand** is defined as a molecule or ion that bonds to a metal ion forming a coordinate (dative covalent) bond by donating a lone pair of electrons into a vacant d-orbital.

A **complex ion** is defined as a central metal ion surrounded by ligands.

Transition metal ions are small and densely charged and can strongly attract electron-rich species called **ligand**s forming **complex ion**s (Figure 34).

Common ligands include $H_2O:$, $:Cl^-$, $:NH_3$ and $:CN^-$, all of which have at least one lone pair of electrons.

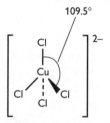

Octahedral (blue)   Tetrahedral (yellow)

Figure 34

OCR(A) A2 Chemistry

In $[Cu(H_2O)_6]^{2+}$, the six electron pairs surrounding the central Cu^{2+} ion repel one another as far apart as possible and the complex ion has an octahedral shape such that all the bond angles are 90°.

In $[CuCl_4]^{2-}$ the four electron pairs surrounding the central Cu^{2+} ion repel one another as far apart as possible and the complex ion has a tetrahedral shape such that all the bond angles are 109°28'.

Coordination number

The **coordination number** of a transition metal, together with the number of lone pairs of electrons, defines the shape of a molecule or ion.

Complex ions with ligands such as H_2O and NH_3 are usually **6** coordinate and **octahedral** in shape. Complex ions with Cl^- ligands are usually **4** coordinate and **tetrahedral** in shape.

Ligands form a dative coordinate bond with a central transition metal ion. Some ligands are able to form two dative coordinate bonds with the central transition metal ion and are known as **bidentate ligands**.

1,2-diaminoethane is a common bidentate ligand (Figure 35). Each N has a lone pair of electrons and each can form a dative bond.

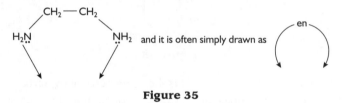

Figure 35

For example, a nickel complex could be drawn as shown in Figure 36.

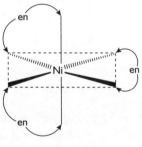

Figure 36

Stereoisomerism in complex ions

Isomerism is commonplace in organic compounds but it can also occur with some inorganic substances.

The square planar structure of $Ni(NH_3)_2Cl_2$ has two different isomeric forms with the ammonia molecules or chloride ions either being on opposite sides of the complex ion (the *trans* form) or alongside each other (the *cis* form) (Figure 37).

Unit F325: Equilibria, Energetics and Elements

> **Examiner tip**
>
> It is easy to lose marks by drawing a complex ion carelessly, showing the coordinate (dative) bond going from the wrong atom. In H_2O, the bond always goes from the O and not the H; in NH_3 the bond always goes from the N.

> The **coordination number** is defined as the total number of coordinate bonds from the ligands to the central transition metal ion in a complex ion.

> **Knowledge check 28**
>
> Deduce the coordination number of the transition element in each of the following:
> (a) $[Zn(NH_3)_4(H_2O)_2]^{2+}$
> (b) $[Ag(NH_3)_2]^+$
> (c) $[Co(NH_3)_5Cl]^{2+}$
> (d) $[Co(H_2NCH_2CH_2NH_2)_3]^{3+}$
> (e) $[Cr(CH_3COO)_2(H_2O)_2]^+$

> **Examiner tip**
>
> Any ion or molecule that has two nitrogens can act as a bidentate ligand and form complexes like the $Ni(en)_3^{2+}$ complex shown in Figure 36. To draw the complex all you have to do is replace 'en' with whatever you are given in the question. Diols and dicarboxylic acids can also behave as bidentate ligands because the oxygen atoms have lone pairs of electrons.

Knowledge check 29

Draw the stereoisomers of $[Co(NH_3)_4Cl_2]^+$ and $[Co(H_2NCH_2CH_2NH_2)_3]^{3+}$. In each case state the shape, the bond angles, the oxidation number of the transition metal ion and the type of isomerism.

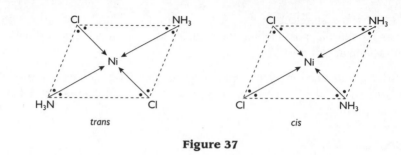

Figure 37

Optical isomerism is also possible in the case of complexes coordinated by polydentate ligands. The nickel 1,2-diaminoethane ion (Figure 38) is an example.

$$en = H_2NCH_2CH_2NH_2$$

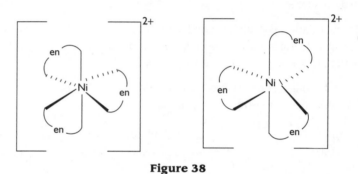

Figure 38

As is the case with organic molecules, it is the asymmetry of the structure that leads to this property and the two molecules shown cannot be superimposed on each other.

Complex ions have a variety of uses but one particularly interesting example is the *cis* form of the molecule $PtCl_2(NH_3)_2$. (The platinum is present as a 2^+ ion and the complex therefore has no overall charge.) The structure is shown in Figure 39.

Figure 39

It is known as *cis*-platin and is used during chemotherapy as an anti-cancer drug. It is a colourless liquid that is usually administered as a drip into a vein and it works by binding to the DNA of cancerous cells and preventing their division. The importance of the exact shape and structure of the molecule is emphasised by the fact that the *trans* molecule is ineffective.

Ligand substitution of complex ions

A ligand substitution reaction takes place when a ligand in a complex ion exchanges for another ligand.

Exchange between H_2O and NH_3 ligands

Water and ammonia ligands have similar sizes such that the coordination number does not change (Figure 40).

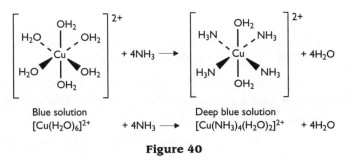

$$[Cu(H_2O)_6]^{2+} \quad + 4NH_3 \longrightarrow \quad [Cu(NH_3)_4(H_2O)_2]^{2+} \quad + 4H_2O$$

Blue solution → Deep blue solution

Figure 40

Exchange between H_2O and Cl^- ligands

Water molecules and chloride ions have *different* sizes and the coordination number changes (Figure 41).

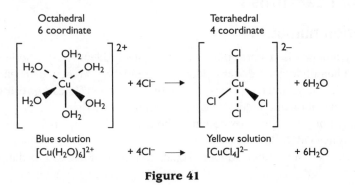

$$[Cu(H_2O)_6]^{2+} \quad + 4Cl^- \longrightarrow \quad [CuCl_4]^{2-} \quad + 6H_2O$$

Blue solution → Yellow solution

Figure 41

A similar reaction takes place when Co^{2+} replaces Cu^{2+} such that $[Co(H_2O)_6]^{2+}$ (which is pink) can form $[CoCl_4]^{2-}$ (which is blue):

$$[Co(H_2O)_6]^{2+} + 4Cl^- \rightarrow [CoCl_4]^{2-} + 6H_2O$$

Stability constants

The strength of binding of a ligand to a cation can be represented quantitatively. When a complex ion such as $[Cu(NH_3)_4(H_2O)_2]^{2+}$ is made it exists in equilibrium with the hydrated copper ion, $[Cu(H_2O)_6]^{2+}$ from which it was made. The equation is:

$$Cu(H_2O)_6{}^{2+} + 4NH_3 \rightleftharpoons Cu(NH_3)_4(H_2O)_2{}^{2+} + 2H_2O$$

Like every other equilibrium, this has an equilibrium constant which in this case is referred to as the stability constant. In this case:

$$K_{stab} = \frac{[Cu(NH_3)_4(H_2O)_2{}^{2+}]}{[Cu(H_2O)_6{}^{2+}][NH_3]^4}$$

K_{stab} has a value of $1.2 \times 10^{13}\,mol^{-4}\,dm^{12}$. There are two important points to note about this equilibrium constant.

Knowledge check 30

Write an equation for each of the following ligand-substitution reactions:

(a) $[Cu(H_2O)_6]^{2+} \rightarrow$ $[Cu(NH_3)_4(H_2O)_2]^{2+}$

(b) $[Co(NH_3)_6]^{3+} \rightarrow$ $[Co(NH_3)_5Cl]^{2+}$

(c) $[Fe(H_2O)_6]^{3+} \rightarrow$ $[Fe(H_2O)_5(SCN)]^{2+}$

- The $2H_2O$ on the right-hand side of the equation are not included in the equilibrium constant expression. This is because the whole reaction is done in water solutions and so the bit extra made as a result of this reaction will make very little difference.
- The square brackets here mean concentration in $mol\,dm^{-3}$ and must not be confused with the brackets that are usually put around the formulae of complex ions to show that they are a complete unit. (To emphasise this, square brackets have been deliberately omitted from the equation.)

The size of this equilibrium constant, $1.2 \times 10^{13}\,mol^{-4}\,dm^{12}$, indicates that the reaction lies well to the right and gives a measure of the greater stability of $[Cu(NH_3)_4(H_2O)_2]^{2+}$ compared to $[Cu(H_2O)_6]^{2+}$.

There is no need to do calculations based on stability constant so long as their significance is understood.

K_{stab} for $[CuCl_4]^{2-}$ is $4.2 \times 10^5\,mol^{-4}\,dm^{12}$. This is a much smaller figure than that of the ammonia complex reflecting the fact that the $[CuCl_4]^{2-}$ complex is not as stable.

Redox reactions

Oxidation number

Oxidation number is a convenient way of quickly identifying whether or not a substance has undergone either oxidation or reduction. Many redox reactions take place in which transition metal ions change their oxidation state by **gaining** or **losing** electrons. Oxidation and reduction can be identified by either:

- movement of electrons — **o**xidation **is** the **l**oss of electrons; **r**eduction **is** the **g**ain of electrons ('OILRIG')
- change in oxidation state/number — oxidation is an increase in oxidation number, reduction is a decrease

The iron(II)–manganate(VII) reaction

The most common redox reaction involving transition elements is the reaction between $Fe^{2+}(aq)$ and $MnO_4^-(aq)$.

Step 1: write a half-ionic equation for each transition metal

In this reaction $Fe^{2+}(aq)$ changes to $Fe^{3+}(aq)$. This can be written as a half-ionic equation:

$$Fe^{2+} \rightarrow Fe^{3+} + 1e^- \qquad \text{Equation (1)}$$

Like any balanced equation the symbols and the charges have to balance. Fe^{2+} loses $1e^-$ when it is **oxidised** to Fe^{3+}.

If Fe^{2+} is oxidised, it follows that $MnO_4^-(aq)$ must be reduced. It is in fact reduced to Mn^{2+}. The first step in constructing a half-equation for this reduction is to recognise the change in oxidation state of the Mn. As oxygen is usually -2, it follows that in MnO_4^- the oxidation state of Mn is $+7$ and for Mn^{2+} the oxidation state of the Mn is $+2$. Therefore Mn changes from $+7$ to $+2$ and hence has to gain $5e^-$:

$$MnO_4^- + 5e^- \rightarrow Mn^{2+}$$

Clearly, this half-equation is *not* balanced. This reaction will not take place unless the MnO_4^- is acidified. Each O in the MnO_4^- forms a water molecule. Since there are 4 oxygens in MnO_4^-, 4 water molecules will be formed and hence 8 H^+ are required:

$MnO_4^- + 8H^+ + 5e^- \rightarrow Mn^{2+} + 4H_2O$ Equation (2)

Each half-equation is now balanced:

$Fe^{2+} \rightarrow Fe^{3+} + 1e^-$ Equation (1)

$MnO_4^- + 8H^+ + 5e^- \rightarrow Mn^{2+} + 4H_2O$ Equation (2)

Step 2: rewrite the half-equations so that the number of electrons in both is the same

In this case, we need to multiply equation (1) by 5 to get:

$5Fe^{2+} \rightarrow 5Fe^{3+} + 5e^-$

and

$MnO_4^- + 8H^+ + 5e^- \rightarrow Mn^{2+} + 4H_2O$ (as before)

Step 3: add the last two half-equations together to cancel out the electrons

So the actual reaction equation is:

$5Fe^{2+} + MnO_4^- + 8H^+ \rightarrow 5Fe^{3+} + Mn^{2+} + 4H_2O$

You can double-check the final equation is correct by making sure the charges on each side balance.

Redox titrations

Transition metal ions are often coloured and the colour changes can be used to show when a titration has reached its end point. The reaction of Fe^{2+} with MnO_4^- is a good example of this. MnO_4^- is purple, while Mn^{2+} is pale pink or colourless. The purple MnO_4^- is added from a burette into acidified Fe^{2+} and it immediately turns pale pink–colourless as the MnO_4^- reacts with the acidified Fe^{2+}. When all of the Fe^{2+} has reacted, the purple colour of the MnO_4^- will remain. The end point of this titration is when a faint permanent pink colour can be seen.

The reaction between Fe^{2+} and MnO_4^- is often tested in the context of a titration calculation.

For example, five iron tablets with a combined mass of 0.900 g were dissolved in acid and made up to $100\,cm^3$ of solution. In a titration, $10.0\,cm^3$ of this solution reacted exactly with $10.4\,cm^3$ of $0.0100\,mol\,dm^{-3}$ potassium manganate(VII). What is the percentage by mass of iron in the tablets?

Step 1: use the balanced equation

$5Fe^{2+}(aq) + MnO_4^-(aq) + 8H^+(aq) \rightarrow 5Fe^{3+}(aq) + Mn^{2+}(aq) + 4H_2O(l)$

Use the balanced equation to obtain the mole ratio of $Fe^{2+}:MnO_4^- = 5:1$. Calculate the number of moles of MnO_4^- by using the concentration and the reacting volume of $KMnO_4$. From the titration results, the amount of $KMnO_4$ can be calculated as:

$$c \times \frac{V}{1000} = 0.0100 \times \frac{10.4}{1000} = 1.04 \times 10^{-4} \text{ mol}$$

From the mole ratio, the amount of Fe^{2+} can be determined:

Fe^{2+} : MnO_4^-

5 : 1

? : 1.04×10^{-4} mol

so, $5 \times 1.04 \times 10^{-4}$ mol Fe^{2+} reacts with 1.04×10^{-4} mol $MnO_4^- = 5.20 \times 10^{-4}$ mol

Step 2: find the amount of Fe^{2+} in the solution prepared from the tablets

10.0 cm^3 of $Fe^{2+}(aq)$ contains 5.20×10^{-4} mol $Fe^{2+}(aq)$. The 100 cm^3 solution of iron tablets contains:

$$10 \times (5.20 \times 10^{-4}) = 5.20 \times 10^{-3} \text{ mol } Fe^{2+}$$

Step 3: Find the percentage of Fe^{2+} in the tablets (A_r: Fe, 55.8)

5.20×10^{-3} mol Fe^{2+} has a mass of $5.20 \times 10^{-3} \times 55.8 = 0.290$ g. Therefore, the % of Fe^{2+} in the tablets is:

$$\frac{\text{mass of } Fe^{2+}}{\text{mass of tablets}} \times 100 \times \frac{0.290}{0.900} \times 100 = 32.2\%$$

You also need to know a second redox titration between iodine and thiosulfate ions:

$$2S_2O_3^{2-}(aq) + I_2(s) \rightarrow S_4O_6^{2-}(aq) + 2I^-(aq) \qquad \text{Equation (1)}$$

This is not usually directly used to determine the concentration of an iodine solution but instead it allows the determination of the concentration of a reagent that generates iodine as a result of a reaction.

An example is the determination of the concentration of a copper sulfate solution. A known volume of copper sulfate is reacted with excess of potassium iodide and the following reaction occurs:

$$2Cu^{2+}(aq) + 4I^-(aq) \rightarrow Cu_2I_2(s) + I_2(s) \qquad \text{Equation (2)}$$

Cu_2I_2 is copper(I) iodide and it forms as a grey–white precipitate. The reaction is therefore a redox process, with the Cu^{2+} being reduced and the I^- being oxidised. The iodine produced is then titrated against a solution of sodium thiosulfate of known concentration.

At the start of the titration, the solution appears brown–purple as a result of the presence of the iodine but as the titration proceeds this colour fades to yellow and the end point is reached when the solution is colourless. The colour change is, in practice, quite hard to see clearly especially as the precipitate of copper(I) iodide tends to get in the way. To help, some starch solution can be added as the end point is approached. This gives a dark blue coloration, which disappears more sharply at the end point.

It can be seen that from:
- equation (2), 2 mol of Cu^{2+} react to produce 1 mol of I_2
- equation (1), the 1 mol of I_2 reacts with 2 mol of $S_2O_3^{2-}$

It follows therefore that for every 1 mol of Cu^{2+}, 1 mol of $S_2O_3^{2-}$ is required. So the amount, in mol, of thiosulfate used is equivalent to the amount, in mol, of copper ion that was present.

> **Knowledge check 31**
>
> 0.426 g of a copper (II) salt is added to 25.0 cm^3 KI(aq). The iodide is in excess. The I_2(aq) produced requires 18.0 cm^3 of a 0.100 mol dm^{-3} solution of $S_2O_3^{2-}$(aq). Calculate the percentage of copper in the original salt. Quote your answer to an appropriate number of significant figures.

Having revised **Module 3: Transition Elements** you should now have an understanding of:
- general properties of transition elements
- precipitation reactions
- ligands and complex ions
- ligand substitution reactions
- redox reactions and titrations

Summary

Questions & Answers

Approaching the unit test

When you finally open the exam paper, it can be quite a stressful moment and you need to be certain of your strategy. The exam paper will consist of a mixture of structured questions and free-response questions. The time allocation for this examination is 2 hours. The total number of marks on the paper is 100. The questions set will relate directly to the Unit F325 content but will also contain a *synoptic* element. Synoptic means that the questions bring together principles and concepts from different areas of chemistry. This means that the questions not only test the content of this unit but also relate to previous units. It is likely that the sections like mole calculations, bonding and energetics from AS Units F321 and F322 will be incorporated into questions.

Time will be tight and you must try to remember to:
- *not* begin to write as soon as you open the paper
- scan *all* the questions before you begin to answer any
- identify those questions about which you feel most confident
- *read the question carefully* — if you are asked to explain, then explain, do not just describe
- take notice of the mark allocation and do not supply the examiner with all your knowledge of any topic if there is only 1 mark allocated (similarly, you will have to come up with *four* ideas if 4 marks are allocated)
- try to stick to the point in your answer (it is easy to stray into related areas that will not score marks and will use up valuable time)
- *try* to answer *all* the questions

Structured questions

These are short-answer questions that may require a single-word answer, a short sentence or a response amounting to several sentences. The setter for the paper will have thought carefully about the amount of space required for the answer and the marks allocated and the space provided usually gives a good indication of the amount of detail required.

Free-response questions

These questions enable you to demonstrate the depth and breadth of your knowledge as well as your ability to communicate chemical ideas in a concise way. These questions often include marks for the quality of written communication (QWC). You are expected to use appropriate scientific terminology and to write in continuous prose, paying particular attention to spelling, punctuation and grammar.

Terms used in the unit test

You will be asked precise questions in the examinations, so you can save a lot of valuable time as well as ensuring you score as many marks as possible by knowing what is expected. Terms most commonly used are explained below.

Define

This requires a precise statement.

Explain

This normally implies that a definition should be given, together with some relevant comment on the significance or context of the term(s) concerned, especially where two or more terms are included in the question. The amount of supplementary comment intended should be interpreted in the light of the indicated mark value.

State

This implies a concise answer with little or no supporting argument.

Describe

This requires students to state in words (using diagrams where appropriate) the main points of the topic. It is often used with reference to mechanisms and requires a step-by-step breakdown of the reaction including, where appropriate, curly arrows to show the movement of electrons. The amount of description intended should be interpreted in the light of the indicated mark value.

Deduce/predict

This implies that students are not expected to produce the required answer by recall but by making a logical connection between other pieces of information. Such information may be wholly given in the question or may depend on answers extracted in an earlier part of the question. Predict also implies a concise answer with no supporting statement required.

Outline

This implies brevity, i.e. restricting the answer to giving essential detail only.

Suggest

This is used in two main contexts. It may either imply that there is no unique answer or that students are expected to apply their general knowledge to a 'novel' situation, one that formally may not be 'in the syllabus'.

Calculate

This is used when a numerical answer is required. In general, working should be shown.

Sketch

When applied to diagrams, this implies that a simple, freehand drawing is acceptable. Nevertheless, care should be taken over proportions and the important details should be clearly labelled.

About this section

This section contains questions similar in style to those you can expect to see in your Unit F325 exam paper. This accounts for 25% of the total marks and the written examination is worth 100 marks. The exam lasts 2 hours and contains synoptic assessment, as well as stretch-and-challenge questions. With synoptic assessment you must, therefore, expect the questions to relate back to other areas of the AS/A2 specification. The stretch-and-challenge questions are designed to test the most able students when awarding the new A* grade.

The questions that follow give you a flavour of the type of questions you will be asked but it is impossible to cover all the topics and all the question styles. A more extensive range of questions are available in the OCR A2 Chemistry textbook published by Philip Allan, which also contains sections on stretch and challenge. Potential A*-grade students should practise answering these questions.

Examiner's comments

Examiner comments on the questions are preceded by the icon ⓔ. They offer tips on what you need to do in order to gain full marks. All student responses are followed by the examiner's comments, indicated by the icon ⓔ, which highlight where credit is due. In the weaker answers, they also point out areas for improvement, specific problems and common errors such as lack of clarity, irrelevance, misinterpretation of the question and mistaken meanings of terms.

Question 1 **The rate equation**

The reaction between hydrogen and nitrogen monoxide is a redox reaction and results in the formation of nitrogen and water.

(a) (i) Write a balanced equation for the reaction.   (1 mark)
(ii) Identify the oxidising agent in the reaction. Justify your answer.   (2 marks)

(b) The rate equation for the reaction is: rate = $k[H_2(g)][NO(g)]^2$.
Using $1.2 \times 10^{-2}\,mol\,dm^{-3}$ $H_2(g)$ and $6.0 \times 10^{-3}\,mol\,dm^{-3}$ NO(g), the initial rate of this reaction was $3.6 \times 10^{-2}\,mol\,dm^{-3}\,s^{-1}$. Calculate the rate constant, k, for this reaction. Quote your answer to 2 significant figures. State the units of the rate constant, k.   (4 marks)

(c) Calculate the initial rate of reaction when each of the following changes is made. Show your working.
(i)   The concentration of the H_2 is tripled.   (1 mark)
(ii) The concentration of the NO is halved.   (1 mark)
(iii) The concentration of both is doubled.   (1 mark)

(d) Dinitrogen pentoxide decomposes according to the equation:
$$2N_2O_5(g) \rightarrow 4NO_2(g) + O_2(g)$$
The decomposition is a first-order reaction with respect to $N_2O_5(g)$.
This decomposition proceeds by a two-step mechanism with the rate-determining step taking place first.
(i)   Write a rate equation for this reaction.   (1 mark)
(ii) Explain the term 'rate-determining step'.   (1 mark)
(iii) Suggest the two steps for this reaction and write their equations.
Show clearly that the two steps equate to the balanced equation given above.   (3 marks)

Total: 15 marks

ⓔ The command word 'calculate' occurs many times in F325 examinations because much of the content is suitable for testing by calculations. In (b) and (c), calculate requires a numerical approach. If more than 1 mark is allocated, as in (b), it is essential to show your working as any errors in the calculation will be marked consequentially. Therefore, provided that you show your working, incorrect answers may score some marks. Do not forget to follow the instructions relating to significant figures.

The command word 'suggest' in part (d) indicates that you have to use your knowledge and apply it to a problem that may not be on the specification.

Student A

(a) (i) $2H_2 + 2NO \rightarrow N_2 + 2H_2O$

Student B

(a) (i) $H_2 + NO \rightarrow \frac{1}{2}N_2 + H_2O$

ⓔ Both students gain the mark. The equation must be balanced and it is acceptable to use fractions.

Student A

(a) (ii) The H_2 has been oxidised because its oxidation number has increased, therefore, the NO must have been the oxidising agent.

Student B

(a) (ii) NO

ⓔ There are only two possible answers, the oxidising agent must be either H_2 or NO, and therefore there is a 50:50 chance of guessing the answer. When questions like this are asked, there are usually no marks for the correct answer. The marks are awarded for the explanation. Both students have given the correct answer but Student B has given no explanation and so fails to score. The explanation given by Student A scores both marks although it is not ideal. It would be better to include relevant oxidation numbers to support the answer. The oxidation number of N in NO is +2 and the oxidation number of N in N_2 is 0. Therefore the reduction in the oxidation number as a result of the reaction identifies NO as the oxidising agent.

A good exam tip is to immediately write the oxidation numbers above each element in the equation:

oxidation numbers	=	0		+2 − 2		0		+1 − 2
		$2H_2$	+	$2NO$	$\rightarrow$	N_2	+	$2H_2O$

so that the changes can be seen easily.

Student A

(b) rate $= k[H_2(g)][NO(g)]^2$

$3.6 \times 10^{-2} = k(1.2 \times 10^{-2})(6.0 \times 10^{-3})^2$

$3.6 \times 10^{-2} = k(1.2 \times 10^{-2})(3.6 \times 10^{-5})$

$3.6 \times 10^{-2} = k(4.32 \times 10^{-7})$

$k = 3.6 \times 10^{-2}/(4.32 \times 10^{-7}) = 83\,333.3 = 8.3 \times 10^4$ to 2 sig figs

Student B

(b) $3.6 \times 10^{-2} = k(1.2 \times 10^{-2})(6.0 \times 10^{-3})^2$

$3.6 \times 10^{-2} = k(1.2 \times 10^{-2})(36 \times 10^{-6})$

$3.6 \times 10^{-2} = k(43.2 \times 10^{-8})$

$k = 3.6 \times 10^{-2}/(43.2 \times 10^{-8}) = 83\,333.3$

ⓔ Both students have calculated the correct numerical value but neither quotes the units for k and both, therefore, lose a mark. Units for the rate constant involve some thought and always carry 1 mark. The question also asks for the answer to 2 significant figures; Student B has ignored this and loses another mark. There is usually a question that tests understanding of significant figures, but by the time you have finished a calculation, it is easy to forget about them.

At A2, it is also possible that a mark might be awarded for the correct use of significant figures without a specific warning being made in the question. If this is the case, there is usually some leeway allowed but it is good practice always to consider the appropriate number of significant

OCR(A) A2 Chemistry

figures to be included in your answer. It is not difficult as all that is required is that the number you give is the same as that of the information that has been supplied. In this particular case you can see that all the data in the question are given to 2 significant figures.

Student A

(c) **(i)** rate triples
    (ii) rate twice as slow
    (iii) rate four times as fast

Student B

(c) **(i)** rate triples
    (ii) rate half as fast
    (iii) rate eight times as fast

e Both students get 1 mark for part (i) but neither gets part (ii) correct. If the concentration of NO is halved, then the rate will change by $(½)^2 = ¼$. Student B gets 1 mark for part (iii). Overall the reaction is third order and if both concentrations are doubled, the rate will change by $(2)^3 = 8$.

Student A

(d) **(i)** rate $= k[N_2O_5]$

Student B

(d) **(i)** rate $= [N_2O_5]$

e Student A gets 1 mark. Student B has forgotten to include the rate constant, k, and loses the mark.

Student A

(d) **(ii)** The slowest step in the mechanism.

Student B

(d) **(ii)** The slowest step.

e Both students get the mark.

Student A

(d) (iii)
$$1N_2O_5 \rightarrow 2NO_2 + O; \quad \text{slow step (RDS)}$$
$$O + 1N_2O_5 \rightarrow 2NO_2 + O_2; \quad \text{fast step}$$
$$2N_2O_5 \rightarrow 4NO_2 + O_2; \quad \text{balanced equation}$$

Student B

(d) (iii)

$$1N_2O_5 \rightarrow N_2O_3 + O_2$$
$$N_2O_3 + N_2O_5 \rightarrow 4NO_2$$
$$2N_2O_5 \rightarrow 4NO_2 + O_2$$

ⓔ Students find devising mechanisms difficult, but both students have given good answers and score full marks. Where a question says 'suggest' it means that the examiner is not expecting a particular answer but is seeing if the student can provide some idea that might be true. The exact mechanism is not relevant, and both students have used the information in the question to suggest valid alternatives.

ⓔ **Overall, Student A scores 12 out of 15, which is grade-A standard. Student B is awarded 9 out of 15 — a grade C. With a little more care and better examination technique, this could easily turn into grade A.**

Question 2 **Equilibrium**

Hydrogen and iodine react according to the equation:

$$H_2(g) + I_2(g) \rightleftharpoons 2HI(g) \qquad \Delta H = +53.0\,kJ\,mol^{-1}$$

(a) State le Chatelier's principle   (1 mark)

(b) Use le Chatelier's principle to predict what happens to the position of the equilibrium when:
    (i)   the temperature is increased
    (ii) the pressure is increased
    (iii) a catalyst is used
    Justify each of your predictions.   (6 marks)

(c) Write an expression for K_c for the equilibrium. State the units, if any.   (2 marks)

(d) (i)   When 0.18 mol of I_2 and 0.5 mol H_2 were placed in a 500 cm³ sealed container
      and allowed to reach equilibrium, the equilibrium mixture was found to contain
      0.010 mol of I_2. Calculate K_c.   (5 marks)
    (ii) Explain what would happen to the value of K_c if the experiment was repeated
      with the 500 cm³ container being replaced by one with a volume of 1 dm³.   (2 marks)

Total: 16 marks

ⓔ The command word 'state' used in part (a) indicates a brief answer is required with no supporting argument; the command word, 'explain' in part (d)(ii) requires a statement and some justification to support that statement. This is also true in (b) where 2 marks are allocated for each sub-part — 1 mark for the prediction and 1 mark for justifying that prediction.

Student A

(a) When a system at equilibrium is subjected to a change, the system will move to try to minimise the effect of the change.

Student B

(a) When a system at equilibrium is subjected to a change in external conditions, the system will move to cancel the effect of the change.

ⓔ Student A scores the mark but Student B does not. Le Chatelier's principle states clearly that the system responds to a change by trying to minimise the effect of that change. It cannot cancel out the change.

Student A

(b) (i) The equilibrium moves to the right because it favours the endothermic forward reaction.
    (ii) No effect because there are an equal number of moles of gas on both sides.
    (iii) By lowering the activation energy the H_2 and I_2 react more readily so that more HI is made. Therefore the equilibrium moves to the right.

Student B

(b) (i) More HI will be produced because heat is absorbed when the forward reaction takes place.
  (ii) Speeds up the reaction because it increases the chance of a collision.
  (iii) Speeds up the reaction by lowering the activation energy.

ⓔ Student A scores 4 out of 6. The prediction about the position of the equilibrium and the explanation in parts (i) and (ii) are correct. What Student A has written in part (iii) is partly correct but the student has failed to realise that the catalyst also affects the reverse reaction such that the position of the equilibrium remains unchanged.

Student B gets 2 marks for part (i) but fails to score for parts (ii) and (iii). The responses in parts (ii) and (iii) relate to the rate of the reaction and not to the position of the equilibrium and therefore no marks can be awarded.

Student A

(c) $K_c = \dfrac{[HI]^2}{[H_2][I_2]}$

There are no units.

Student B

(c) $K_c = \dfrac{[HI]^2}{[H_2][I_2]}$

(no units)

ⓔ Both students get 2 marks.

Student A

(d) (i)

	$H_2(g)$	+	$I_2(g$	$\rightleftharpoons$	$2HI(g)$
Initial mol	0.5		0.18		0
Final mol			0.010		

It follows that 0.17 mol of I_2 and therefore 0.17 mol of H_2 also reacted. Hence the mol of H_2 left is 0.5 − 0.17 = 0.33 mol.
For each mol of I_2 and H_2 that react 2 mol of HI are formed, therefore the equilibrium amount of $2 \times 0.17 = 0.34$ mol. K_c is measured in terms of concentration therefore each of the equilibrium amounts must be converted to $mol\,dm^{-3}$
$I_2 = 0.01/0.5 = 0.02\,mol\,dm^{-3}$
$H_2 = 0.33/0.5 = 0.66\,mol\,dm^{-3}$
$HI = 0.34/0.5 = 0.68\,mol\,dm^{-3}$
$K_c = (0.68)^2/(0.02)(0.66) = 35$

Student B

(d) (i) $K_c = \dfrac{[HI]^2}{[H_2][I_2]}$

$[I_2] = 0.01$ mol and $[H_2] = 0.5 - 0.17 = 0.33$

$[HI] = 0.34$

$K_c = 0.34/(0.01 \times 0.33) = 103$

e Student A has worked through the calculation systematically, obtained the correct answer and scores 5 marks. Student B has made progress but carelessness once again means that marks are lost. First, the use of '[ ]' indicates that a concentration is being stated. However, in this case the student is referring to an amount in mol so statements such as $[I_2] = 0.01$ mol must not be used. Student A wisely converts the amounts in mol into concentrations before calculating the equilibrium constant. Student B does not do so and also puts 0.34 instead of $(0.34)^2$. In fact, in this case, since the total number of particles does not change as a result of the reaction, the volume of the container does not matter and the correct answer would be obtained using the amount in moles. It is not clear that Student B understands this. The result is that no more than 3 marks can be given.

Student A

(d) (ii) If the volume of the container is increased to $1\,dm^3$, the equilibrium constant must be calculated as:

$I_2 = 0.01/1 = 0.01\,mol\,dm^{-3}$

$H_2 = 0.33/1 = 0.33\,mol\,dm^{-3}$

$HI = 0.34/1 = 0.34\,mol\,dm^{-3}$

$K_c = (0.34)^2/(0.01)(0.33) = 35$

Which seems to be the same as before.

Student B

(d) (ii) K_c will be half what it was $= 103/2 = 51.5$.

e Student A seems rather surprised by the result and perhaps does not fully understand why but nonetheless the response is correct and 2 marks are awarded.

Student B seems quite confused and the answer is perhaps just a guess. No marks can be given.

e **Overall, Student A scores 14 out of 16 — a good mark. Student B has not revised this topic carefully enough and only scores 7, which is no more than D/E-grade standard.**

Question 3 pH

(a) (i) A weak organic acid, HA, has the percentage composition by mass: C, 40%; H, 6.7%; O, 53.3%. Calculate the empirical formula of HA. (2 marks)

   (ii) HA has a relative molecular mass of 60.0. What is its molecular formula? (1 mark)

(b) 1.20 g of HA was dissolved in 250.0 cm³ water. Calculate the pH of the resulting solution. Show all of your working. (K_a of HA = 1.7×10^{-5} mol dm^{-3}) (5 marks)

(c) A 0.04 mol dm^{-3} solution of HA was titrated with a 0.05 mol dm^{-3} sodium hydroxide solution.
   (i) Calculate the pH of the NaOH(aq). ($K_w = 1.0 \times 10^{-14}$ mol² dm^{-6}) (2 marks)
   (ii) Calculate the volume of NaOH(aq) required to neutralise 25.0 cm³ of solution HA. (3 marks)
   (iii) Sketch a graph to show the change in pH during the titration. (4 marks)

(d) Indicators can be used to determine the end point of a titration. Which of the following would be most suitable for this titration? Justify your answer and suggest what you would see at the end point. (3 marks)

Indicator	Acid colour	pH range	Alkaline colour
Thymol blue (acid)	Red	1.2–2.8	Yellow
Bromocresol purple	Yellow	5.2–6.8	Purple
Thymol blue (base)	Yellow	8.0–9.6	Blue

Total: 20 marks

ⓔ When asked to sketch a graph as in part (c) a great deal of care is required and it is important to label the axes. Parts (b) and (c)(i) require calculation of the pH of the acid and base respectively; (c)(ii) requires calculation of the volume of base required to neutralise the acid. Each of these factors has to be taken into account when sketching the graph.

Student A

(a) (i) Empirical formula:
   C : H : O
   40/12.0 : 6.7/1.0 : 53.3/16.0
   3.33 : 6.7 : 3.33
   1 : 2 : 1 = CH_2O
   (ii) CH_2O has a mass = 12.0 + 2.0 + 16.0 = 30.0
   therefore empirical mass × 2 = molecular mass
   ∴ molecular formula = $C_2H_4O_2$

Student B

(a) (i) Molecular mass is 60.0
   C is 40% = 24 = 2C
   H is 6.7% = 4.02 = 4H
   O is 53.3% = 31.98 = 2O
   So formula is $C_2H_4O_2$ which means empirical formula is CH_2O.
   (ii) Molecular formula is $C_2H_4O_2$.

@ Student A gets 3 marks. Student B has not answered the question in the order presented. The relative molecular mass is given in part (ii) but has been used by the student to answer part (i). This is not the correct way of handling this question and it could be that the student would gain only the mark for part (ii) (i.e. I). The examiner *might* allow the marks for part (i) but it is certainly not a wise way of tackling this question.

Student A

(b) $K_a = \dfrac{[H^+][A^-]}{[HA]} = \dfrac{[H^+]^2}{HA}$

$\therefore [H^+]^2 = K_a \times [HA]$

$K_a = 1.7 \times 10^{-5}$, $[HA] = 1.2/60 = 0.02$

$\therefore [H^+]^2 = 1.7 \times 10^{-5} \times 0.02 = 3.4 \times 10^{-7}$

$H^+ = \sqrt{3.4 \times 10^{-7}} = 5.8 \times 10^{-4}$

$pH = -\log_{10}[H^+] = -\log_{10}(5.8 \times 10^{-4}) = 3.23$

Student B

(b) $pH = -\log_{10}\sqrt{K_a \times [HA]}$

$pH = 4.69$

@ Such questions are difficult to mark because there are a number of valid ways of carrying out the calculation. The instructions ask the students to 'Show all of their working.' Neither of the students has the correct answer of pH = 2.93. Student A scores 4 out of 5; Student B scores only I mark.

Student A has shown all the working and it is therefore possible to see where any mistakes have been made. The only error made by Student A is in working out the concentration of HA. By using the value 1.2/60.0 = 0.02 the number of moles of HA in 250 cm^3 has been worked out but not the *concentration* of HA. The correct concentration is 0.08 mol dm^{-3}.

It is impossible to deduce where Student B has gone wrong and so the only mark that can be awarded is for quoting an equation, $pH = -\log_{10}\sqrt{K_a \times [HA]}$, which could be used to obtain the correct answer.

Try the calculation: remember $K_a = 1.7 \times 10^{-5}$ and [HA] =0.08.

Student A

(c) (i) $K_w = [H^+][OH^-] = 1.0 \times 10^{-14}$

$[H^+][0.05] = 1.0 \times 10^{-14}$

$\therefore [H^+] = 1.0 \times 10^{-14}/0.05 = 2.0 \times 10^{-13}$

$pH = -\log_{10}[H^+] = -\log_{10} 2.0 \times 10^{-13} = 12.7$

Student B

(c) (i) $pOH = -\log_{10}[OH^-] = -\log_{10}(0.05) = 1.30$

$pH = 14 - pOH = 14 - 1.30 = 12.7$

e Both students score 2 marks. The methods adopted are different but both are valid and lead to the correct answer.

Student A

(c) (ii) HA   +   NaOH   →   NaA   + H_2O

1 mol   :   1 mol

moles of HA = moles of NaOH , $n = cV = 0.04 \times 25.0/1000 = 0.001$

volume of NaOH = $V = n/c = 0.001/0.05 = 0.02\,dm^3 = 20\,cm^3$

Student B

(c) (ii) $\dfrac{0.04 \times 25.0}{0.05} = 20\,cm^3$

e Both students get all 3 marks but Student B is living dangerously by providing no explanation. It is vital to show your working in any calculation.

Student A

(c) (iii)

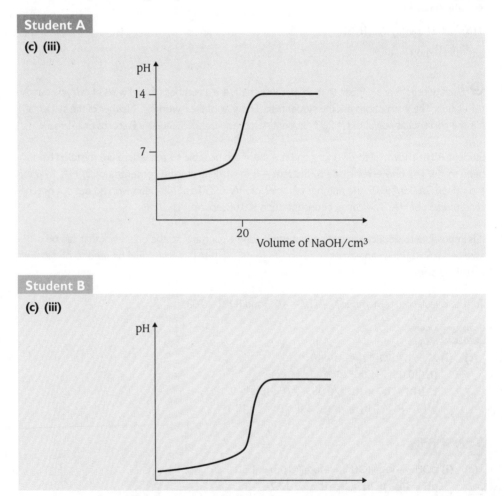

Student B

(c) (iii)

ⓔ The marking points for the sketch would be: correct labels including units, initial and final pH approximately correct, correct shape, rapid change in pH from about 7–12 after the addition of $20\,cm^3$ of NaOH.

Student A has been methodical and gains all 4 marks. Student B also understands the chemistry but only scores 2 marks. Use the mark scheme above and see if you can identify where Student B has lost 2 marks.

Student A

(d) Thymol blue (base) would be the best indicator because it changes colour in the pH region 8.0–9.6, which matches the rapid change in pH for this reaction.

Student B

(d) Thymol blue (base) would be the best indicator because it changes colour in the pH region 8.0–9.6. The end point would be a green colour.

ⓔ Both students have selected the correct indicator. Student A has given a good explanation of the choice of thymol blue (base) but has forgotten to say what would be seen at the end point. Student B has not explained why the indicator has been selected and has simply copied the pH range from the question. The end point should occur when there is an equal amount of the acid and the alkaline forms of the indicator, i.e. an equal amount of yellow and blue, hence the end point would be green. Both students score 2 marks.

ⓔ **Both students seem to understand the chemistry, but Student B has failed to use this well. Student A scores 17 out of 19; Student B only obtains 11. The net result of poor technique is that Student B is underachieving by about two or three grades. Look back carefully at Student B's responses and identify where marks could easily have been gained.**

Question 4 **Acids and bases**

A patient suffering from a duodenal ulcer displays increased acidity in their gastric juices. The exact acidity of the patient's gastric juice is monitored by measuring the pH.

(a) (i) Define pH. (1 mark)
   (ii) On one day the patient's gastric juice was found to have a hydrochloric acid concentration of $8.0 \times 10^{-2}\,mol\,dm^{-3}$.
Calculate the pH of the gastric juice. (1 mark)

(b) One of the most common medications designed for the relief of excess stomach acidity is aluminium hydroxide, $Al(OH)_3$.
   (i) Write an equation for the reaction between HCl and $Al(OH)_3$. (1 mark)
   (ii) On another day the gastric juice of the patient was found to have a pH of 1.3. The patient produces $2\,dm^3$ of gastric juice in a day. This volume of gastric juice is to be treated with tablets containing $Al(OH)_3$ to raise the pH to 2.
Calculate the mass of aluminium hydroxide required to raise the pH of $2\,dm^3$ of gastric juice from 1.3 to 2. (5 marks)

(c) The control of the pH of blood is also important. This is achieved by the presence of HCO_3^- ions in blood plasma. Using appropriate equations, explain how HCO_3^- can act as a buffer solution. (3 marks)

Total: 11 marks

e The command word 'calculate' in parts (a)(ii) and (b)(ii) requires a numerical approach. In (b)(ii) it is important that you show your working as any errors in the calculation will be marked consequentially. The calculation accounts for almost half the marks in this question. Part (c) requires knowledge of the equilibrium involved along with an explanation of how the equilibrium counters changes in acidity and alkalinity.

Student A

(a) (i) pH is minus the logarithm to base 10 of the hydrogen ion concentration in a solution.

Student B

(a) (i) pH = $-\log[H^+]$

e Both students get the mark. Student A has made it harder to answer the question by defining pH in words. Some students seem to think that definitions should always be given in words. This is not the case and Student B has the better approach.

Student A

(a) (ii) pH = $-\log_{10}(8.0 \times 10^{-2}) = 1.10$

Student B

(a) (ii) pH = 1.097

ⓔ Both students get the mark although both quoted their answers to too many significant figures. If the question had carried 2 marks, Student B, in particular, might have only scored 1.

Student A

(b) (i) $3HCl + Al(OH)_3 \rightarrow AlCl_3 + 3H_2O$

Student B

(b) (i) $3HCl + Al(OH)_3 \rightarrow AlCl_3 + 3H_2O$

ⓔ Both students get the mark.

Student A

(b) (ii) A pH of 2 means $[H^+] = 10^{-2}$ or $0.01\,mol\,dm^{-3}$
A pH of 1.3 means $[H^+] = 10^{-1.3} = 0.05\,mol\,dm^{-3}$
The pH has to be raised by $0.04\,mol$
$3HCl + Al(OH)_3 \rightarrow AlCl_3 + 3H_2O$
Therefore $0.04/3\,mol$ of $Al(OH)_3$ are needed $= 0.0133\,mol$
$1\,mol$ of $Al(OH)_3 = 78.0\,g$
So $Al(OH)_3$ needed is $0.0133 \times 78.0 = 1.04\,g$

Student B

(b) (ii) A pH of 2 is $0.01\,mol\,dm^{-3}$
A pH of 1.3 is $0.02\,mol\,dm^{-3}$
So $2\,dm^3$ of gastric juice has $0.04\,mol$ and must become 0.02 by adding $Al(OH)_3$
Using the equation gives mol of $Al(OH)_3$ of $0.02/3$
$Al(OH)_3 = 78.0\,g$
So $Al(OH)_3$ needed is $78.0 \times 0.02/3 = 0.52\,g$

ⓔ This is quite a difficult question and is perhaps intended as one of the stretch-and-challenge parts that are included to test the best students. Although both students have obtained the wrong answer both have done well and achieved 4 out of the 5 marks.

In the mark scheme there is a mark each for converting the two pH values to concentrations of H^+. There is 1 mark is for working out the amount, in mol, of H^+ that has to be neutralised. The fourth mark is for the moles of aluminium hydroxide required and the final mark for converting this into grams.

Student A forgets that the volume being treated is $2\,dm^3$ and so concludes that $0.04\,mol$ of HCl must be neutralised by the $Al(OH)_3$. The figure should be $0.08\,mol$ for the $2\,dm^3$. So the third mark is lost. After that the calculation is correct, so only 1 mark is dropped overall. The examiner will always award marks for later parts of a question if they are carried out correctly even if an error has been made earlier. Student A makes a statement that is worded poorly. It is incorrect to write 'The pH has to be raised by 0.04 mol' instead of 'To raise the pH from 1.3 to 2 requires the amount, in moles, of H^+ to be changed from 0.05 to 0.01' but in a calculation the examiner is unlikely to be too hard on poor wording and it is better not to worry too much about it. Be careful though always to provide correct units for quantities as these might carry marks.

Student B fails to convert the pH of 1.3 to the correct concentration of hydrogen ions. This is a common error and often occurs because students have not learned how to use their calculators reliably. You should make certain this does not happen to you. Although again rather poorly worded, this student does remember that $2\,dm^3$ is being used and from this stage proceeds to complete the calculation correctly.

Student A

(c) The addition of H^+ causes the HCO_3^- ion to convert to CO_2
$$HCO_3^- + H^+ \rightleftharpoons CO_2 + H_2O$$
So this allows any excess acid to be partly controlled
The addition of alkali converts the HCO_3^- to CO_3^{2-}
$$HCO_3^- + OH^- \rightleftharpoons CO_3^{2-} + H_2O$$
This is not very likely in the blood though

Student B

(c) The buffer works like
$$HCO_3^- + H^+ \rightleftharpoons CO_2 + H_2O$$
$$HCO_3^- + OH^- \rightleftharpoons CO_3^- + H_2O$$

ⓔ Student A provides a satisfactory answer and, in particular, does well to mention that buffers only partly control the pH. A common error is to say that buffer solutions stop the pH from changing. This would always lose a mark as, although buffer solutions stabilise pH well, they cannot totally stop the pH from changing. Student A scores 3 marks.

Student B does not give any explanation and therefore loses 1 mark. The first equation is correct but the second has the carbonate ion as CO_3^-, which is a serious error at A2 and means a second mark is lost. Therefore only 1 mark is obtained for this part of the question.

ⓔ **Overall, Student A scores 10 out of 11, an excellent mark. Student B gets 8 out of 11, which falls just short of grade-A standard because of careless errors.**

Question 5 **Born–Haber cycles**

(a) (i) Explain what is meant by the term lattice enthalpy. (2 marks)

    (ii) Write an equation to show what is meant by the lattice enthalpy of magnesium chloride. (2 marks)

(b) Use the Born–Haber cycle below to answer the questions that follow.

Energy/kJ mol⁻¹

$Mg^{2+}(g) + 2e^- + 2Cl(g)$

$D = +244$

$E = -698$

$Mg^{2+}(g) + 2e^- + Cl_2(g)$ $Mg^{2+}(g) + 2Cl^-(g)$

$C = +1450$

$Mg^+(g) + 1e^- + Cl_2(g)$

$B = +736$ $F = -2522$

$Mg(g) + Cl_2(g)$

$A = +148$

$Mg(s) + Cl_2(g)$ G $MgCl_2(s)$

    (i) Identify which step represents the second ionisation energy of magnesium. (1 mark)

    (ii) Write an equation that illustrates the second ionisation energy of magnesium. (1 mark)

    (iii) Explain why the enthalpy value for the second ionisation energy of magnesium is about twice the value of the first ionisation energy of magnesium. (2 marks)

    (iv) Write an equation that illustrates the first electron affinity of chlorine. (2 marks)

    (v) State the energy, in kJ mol⁻¹, for the first electron affinity of chlorine. (1 mark)

    (vi) Calculate the enthalpy of formation of magnesium chloride. (2 marks)

(c) The lattice enthalpy for magnesium bromide is −2440 kJ mol⁻¹.
Explain the difference in the values of the lattice enthalpies of magnesium bromide and magnesium chloride. (1 mark)

(d) Magnesium bromide is soluble in water.
The enthalpy of hydration of magnesium ions is −1921 kJ mol⁻¹ and the enthalpy of hydration of the bromide ion is −336 kJ mol⁻¹.
Calculate the enthalpy of solution of magnesium bromide. (2 marks)

(e) Using enthalpies of solution is not always a reliable way of predicting whether a substance will be soluble in water.

    (i) What other energy change should be considered in making a prediction of solubility? (1 mark)

    (ii) Explain whether this other energy change is likely to suggest that a substance will be more or less soluble in water. (2 marks)

(f) Describe how you could distinguish between aqueous solutions of magnesium bromide and magnesium chloride. State the observations you would make. (3 marks)

Total: 22 marks

ⓔ There are several parts in this question and each starts with a command word. It is important that you follow the instructions precisely. 'State' requires a brief answer with no supporting argument; 'explain' and 'describe' require more detailed answers with some additional comment.

Student A

(a) (i) The lattice enthalpy is the enthalpy change when 1 mol of ionic solid is formed from its ions in the gas state.

Student B

(a) (i) It is the enthalpy released when the constituent gaseous ions form 1 mol of ionic solid.

ⓔ Both score 2 marks. It is essential that you learn straightforward definitions. The marks in this definition are for: forming 1 mol of ionic solid ✔, from the gaseous ions ✔.

Student A

(a) (ii) $Mg^{2+}(g) + 2Cl^-(g) \rightarrow MgCl_2(s)$

Student B

(a) (ii) $Mg^{2+} + Cl_2^-(g) \rightarrow MgCl_2(s)$

ⓔ Student A gets both marks but Student B loses both marks.

Student B has not included the state symbol, (g), after Mg^{2+}. Many students forget to put state symbols in equations. In cases such as this, where a change in state is involved, they are essential. Do not be caught out by this.

Student B has also made another common error: $2Cl^-(g)$ is not the same as $Cl_2^-(g)$. Writing $Cl_2^-(g)$ implies that the two Cl are joined together and have made a Cl_2 molecule which has then received a '−' charge. This is incorrect; each Cl has received a '−' charge and so this must be written as $2Cl^-$.

Student A

(b) (i) C

Student B

(b) (i) C

ⓔ Both students get the mark.

Student A

(b) (ii) $Mg^+(g) \rightarrow Mg^{2+}(g) + 1e^-$

OCR(A) A2 Chemistry

Student B

(b) (ii) $Mg(s) \rightarrow Mg^{2+}(g) + 2e^-$

ⓔ Student A gets both marks and has made good use of the information in the question. Student B fails to score.

The equation given by Student B is incorrect on a number of counts. Most important is that the equation shows the first and second ionisation energies combined together. In addition, Student B has made another careless error in thinking of magnesium as a solid and adding this as the state symbol in the equation. He/she has forgotten that for the process of ionisation the magnesium has first been atomised and is therefore a gas.

Student A

(b) (iii) Mg^+ is smaller than the Mg atom and therefore it is more difficult to remove the second electron.

Student B

(b) (iii) The second ionisation energy removes two electrons but the first only one, therefore it is twice as much.

ⓔ This is a difficult concept. The ease with which an electron can be removed depends on the attraction between the protons in the nucleus and the outer electrons. The three main factors that influence this are:
- distance from the nucleus
- shielding by inner shells
- proton to electron ratio

You must remember to consider all three factors to identify which of them might be relevant to the answer expected. The shielding remains constant for the first and second ionisation energies of Mg, but the distance from the nucleus and the proton:electron ratio both change. Student A scores 1 mark for explaining the variation in size/distance from the nucleus, but Student B fails to score. Student B has fallen for the trap set by the question and jumped to the wrong conclusion.

Student A

(b) (iv) $Cl(g) + 1e^- \rightarrow Cl^-(g)$

Student B

(b) (iv) $Cl(g) + e^- \rightarrow Cl^-(g)$

ⓔ Both students get the mark.

Student A

(b) (v) $-698 \, kJ \, mol^{-1}$

Student B

(b) (v) 349

ⓔ Neither student gets the mark. Student A has used the information in the question but failed to spot that the enthalpy change given is for 2 × first electron affinity of chlorine. Student B has spotted this and has halved the numerical value but has forgotten to put in the negative sign and also has omitted to put units. Student B, therefore, also fails to score.

Student A

(b) (vi) $\Delta H_f = 148 + 736 + 1450 - 698 - 2522 = -642\,kJ\,mol^{-1}$

Student B

(b) (vi) $-642\,kJ\,mol^{-1}$

ⓔ Both students get 2 marks but Student A shows better examination technique by showing the working.

Although you should use a calculator to avoid making an error in the arithmetic, it is wise to write down your working. If you then make a mistake, the examiner can check whether a mark can still be awarded. It is surprising how often this turns out to be the case. Although a correct answer will normally score full marks, of course no mark will be awarded if there is an error and no working is shown. In some cases a question will specify that working *must* be shown and in this case marks are awarded specifically for the steps of the calculation.

Student A

(c) The lattice enthalpy depends on the ionic radius and the size of the charge. The charge for Cl^- and Br^- is the same but the Cl^- is smaller than the Br^- and therefore the attraction between the Mg^{2+} ion and the Cl^- will be greater.

Student B

(c) Charge density of $Cl^- > Br^-$
∴ lattice enthalpy $MgCl_2$ is bigger than $MgBr_2$

ⓔ Student A would definitely get the mark. Student B would probably score too, but he/she has not given a clear answer to the question. The fault lies in the use of the word 'bigger'. This is arguable because with lattice energies we are dealing with negative numbers. It is true that the numerical value of the lattice energy is greater for magnesium chloride than it is for magnesium bromide but is -2522 bigger or smaller than -2440? In answering questions of this type where numbers are negative it is best to avoid saying that the lattice energy is bigger, greater, more, less and smaller and say instead that the numerical value of the lattice energy is greater or smaller or that it is more negative.

Student A

(d)

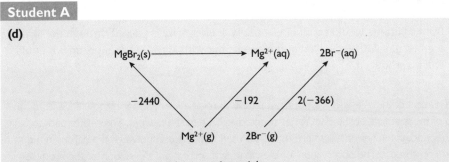

$$\Delta H_1 = 2440 - 1921 - (2 \times 336) = -153\,kJ\,mol^{-1}$$

Student B

(d) Enthalpy of solution = $-2440 + 1921 + 772 = 153\,kJ\,mol^{-1}$

e Student A gets both marks but Student B has made an error in the sign of the answer and therefore scores only 1 mark. In questions involving enthalpy changes it is always a good idea to provide a diagram indicating the direction of the enthalpy changes involved. It helps to avoid unnecessary errors.

Student A

(e) (i) The entropy change

Student B

(e) (i) Entropy

e Both students get the mark.

Student A

(e) (ii) Entropy increases when a substance dissolves in water as the particles can move more readily. This means that substances will be predicted to be more soluble than when just enthalpy is used.

Student B

(e) (ii) More soluble

e Student A scores the 2 marks but Student B fails to score as no explanation is provided.

If a question is asked to which the answer is 50:50 (i.e. in this case it could either be 'more' or 'less') no mark is available without an explanation.

Student A

(f) Add silver nitrate to each and observe the colour of the precipitate. $MgCl_2$ would give a white solid and $MgBr_2$ would give a yellow solid.

Student B

(f) Both solutions would conduct electricity. If electricity is passed through the $MgCl_2$ a green gas will be evolved at the anode, but with $MgBr_2$ an orange/brown liquid will be produced.

🄮 This part of the question is synoptic: you have to recall information and knowledge from other areas of the specification. Student A has selected $AgNO_3$ as the reagent. This relates back to Unit F321 (Foundation Chemistry), where $AgNO_3$ is used to distinguish between the halides in the section on group 7. The observation for the chloride is correct but the colour for the bromide is incorrect. The iodide gives a yellow precipitate with $AgNO_3$ and the bromide gives a cream precipitate. Student A, therefore, gets 2 marks and perhaps is quite lucky not to lose a further mark for not specifying that the silver nitrate should be aqueous. When describing chemical tests you should always consider the state of the reagents as this could be included in the mark scheme.

Student B has given an unexpected answer, but nevertheless it is good chemistry and would probably work. Student B might get all 3 marks although it would be difficult to detect the chlorine by colour.

🄮 **Student A shows a good grasp of this topic and has been systematic in supplying the answers. Good use is made of the information in the question. The score of 19/22 is comfortably grade-A standard.**

Student B scores 12/22 — equivalent to a C grade — with marks lost mostly by carelessness.

Question 6 **Enthalpy, entropy and free energy**

(a) Use the data below to calculate the standard enthalpy change for the reaction:

$C(s) + CO_2(g) \rightarrow 2CO(g)$

$\Delta H_f^{\ominus}(CO_2) = -393 \text{ kJ mol}^{-1}$

$\Delta H_f^{\ominus}(CO) = -110.5 \text{ kJ mol}^{-1}$

(2 marks)

(b) Use the data below to calculate the standard entropy change for the reaction:

$C(s) + CO_2(g) \rightarrow 2CO(g)$

$S^{\ominus}(CO_2) = 213.8 \text{ J mol}^{-1} \text{ K}^{-1}$; $S^{\ominus}(C) = 5.7 \text{ J mol}^{-1} \text{ K}^{-1}$;

$S^{\ominus}(CO) = 197.9 \text{ J mol}^{-1} \text{ K}^{-1}$

(2 marks)

(c) (i) State the relationship between ΔG, ΔH and ΔS.

(1 mark)

   (ii) Use your answers to (a) and (b) above to determine the value of $\Delta G^{\ominus}$ for the reaction of carbon dioxide and carbon under standard conditions of 298 K and 101 kPa.

(2 marks)

(d) Calculate the minimum temperature in °C required for the reaction between carbon dioxide and carbon to become feasible.

(3 marks)

Total: 10 marks

 The command word 'calculate' features in parts (a), (b) and (d) and requires a numerical approach. In part (c)(ii) 'determine' also requires a numerical approach. Therefore 9 of the 10 marks in this question involve calculations.

Student A

(a)

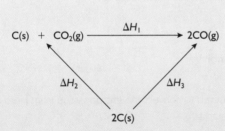

$\Delta H_1 = -\Delta H_2 + \Delta H_3 = 393.7 - (2 \times 110.5) = 172.7 \text{ kJ mol}^{-1}$

Student B

(a) Enthalpy change $= -(2 \times 110.5) - (-393.7) = -172.7 \text{ kJ mol}^{-1}$

 Although enthalpy cycles were studied in AS Unit F322, they are clearly relevant to the study of free energies in A2 Unit F325. You must therefore be ready to complete calculations of this type.

Student A gets both marks. It is not a requirement that an enthalpy diagram is provided but it may help to avoid errors. Student B sees the enthalpy change as the (enthalpy of the products) − (enthalpy of the reactants), which is fine but the student has become muddled working out the answer and has ended up with the wrong sign for the answer. Only 1 mark is scored. It is all too easy to make this mistake and all calculations should be checked carefully. It is always worthwhile to look at the reaction and consider whether you expect it to be exothermic or endothermic.

Student A

(b) The entropy change is $2S^{\ominus}(CO) - S^{\ominus}(CO_2) - S^{\ominus}(C) = 2 \times 197.9 - 213.8 - 5.7 = 176.3\,kJ\,mol^{-1}$

Student B

(b) Change is $2 \times 197.9 - 213.8 - 5.7 = 176.3\,J\,mol^{-1}\,K^{-1}$

ⓔ Both students have carried out the arithmetic correctly but Student A has incorrectly written the units as $kJ\,mol^{-1}$. Therefore Student A scores only 1 mark but Student B scores 2 marks.

Student A

(c) **(i)** $\Delta G = \Delta H - T\Delta S$

Student B

(c) **(i)** $\Delta G = \Delta H - T\Delta S$

ⓔ Both students get the mark.

Student A

(c) **(ii)** $\Delta G = 172.7 - 298 \times 0.1763 = 120.2\,kJ\,mol^{-1}$

Student B

(c) **(ii)** $\Delta G = -225.2\,kJ\,mol^{-1}$

ⓔ Student A has remembered to convert the entropy value from J into kJ and has completed the calculation correctly for 2 marks.

Student B has used the wrong answer obtained in part (a) and has then completed the calculation correctly. In general, the examiner allows marks for calculations that are incorrect if they are based on an error in a previous part of the question which has already been penalised. If Student B had shown the working of the calculation, 2 marks would have been awarded. However, if the examiner cannot see how the answer has been obtained, these marks may be lost.

Student A

(d) For a reaction to be feasible $\Delta G = 0$, so $\Delta H = T\Delta S$
$172.7 = T(0.1763)$
$T = 979°C$

OCR(A) A2 Chemistry

Student B

(d) At equilibrium $\Delta H = T\Delta S$

$-172.7 = 0.1763T$

$T = -979$ or $-979 - 273 = -1252°C$

ⓔ Student A scores 2 out of the 3 marks as the answer given is in K and not °C. The correct answer is $979 - 273 = 706°C$.

Student B is still suffering from the error made in part (a) but this time has shown the working of the calculation. It is all correct and, even though the answer is wrong, the examiner will award 3 marks. It is a pity that the student, once a clearly wrong answer had been obtained, did not check the arithmetic. However this may not have been possible if time was short. Sometimes students are tempted to cross out parts of questions where they have recognised that the answer cannot be right. This is a mistake because, as in this case, it might still be possible to get some credit for what is there.

ⓔ **Student A has dropped 2 marks and obtained 8 out of 10, which is grade-A standard, but it could have been full marks with just a little more care. Student B would perhaps have only scored 7. It is possible that it might be more if the examiner recognised what had happened in part (c)(ii) but this cannot be guaranteed.**

Question 7 Redox equations and electrode potentials

(a) Draw a diagram to show how the standard electrode potential of the half-cell $Fe^{3+}(aq) + e^- \rightleftharpoons Fe^{2+}(aq)$ would be measured. State the conditions necessary. (6 marks)

(b) Use the following electrode potentials to predict whether, under standard conditions, $Fe^{3+}(aq)$ will be able to react with:

(i) $I^-(aq)$

(ii) $Br^-(aq)$

$I_2(aq) + 2e^- \rightleftharpoons 2I^-(aq)$	$E^\ominus = +0.54\,V$
$Fe^{3+}(aq) + e^- \rightleftharpoons Fe^{2+}(aq)$	$E^\ominus = +0.77\,V$
$Br_2(aq) + 2e^- \rightleftharpoons 2Br^-(aq)$	$E^\ominus = +1.09\,V$

If a reaction is possible, state what you would observe as the reaction took place. (5 marks)

(c) $I^-(aq)$ reacts with acidified $KMnO_4$ to form $Mn^{2+}(aq)$ ions and $I_2(aq)$.
Write half-equations for each of these reagents and use them to construct a balanced ionic equation for the reaction. (3 marks)

Total: 14 marks

ⓔ Part (a) requires a diagram but care is needed as 6 marks are available. Before starting, plan what you need to include for the 6 marks — the hydrogen electrode ✔, the Fe^{2+}/Fe^{3+} electrode ✔, the salt bridge and the voltmeter ✔ as well as the conditions of temperature ✔, pressure ✔ and concentrations ✔ are all required.

Student A

(a)

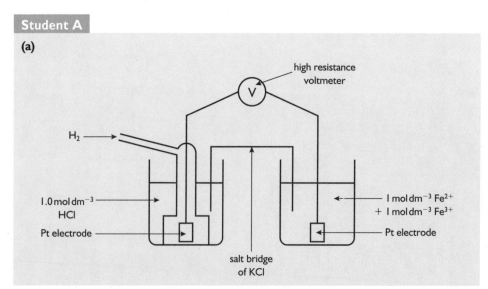

OCR(A) A2 Chemistry

Student B

(a)

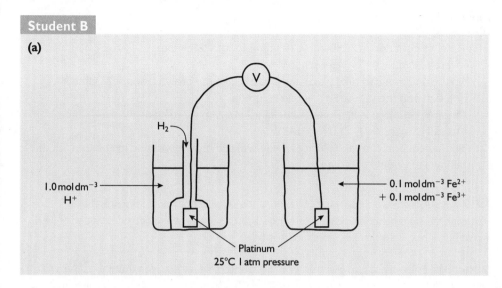

H_2

$1.0\,mol\,dm^{-3}$
H^+

$0.1\,mol\,dm^{-3}\ Fe^{2+}$
$+ 0.1\,mol\,dm^{-3}\ Fe^{3+}$

Platinum
25°C 1 atm pressure

e Student A has learnt this topic well and scores all 6 marks.

Student B knows the work quite well but has failed to include a salt bridge in the diagram. This loses 1 mark. The diagram is drawn poorly but so long as it is clear this will not lose marks. A good point about Student B's work is that the concentrations chosen for the $Fe^{2+}(aq)$ and $Fe^{3+}(aq)$ are more realistic than those of Student A. The value of the electrode potential measured will be the same for redox pairs such as $Fe^{2+}(aq)$ and $Fe^{3+}(aq)$ if their concentrations in $mol\,dm^{-3}$ are the same. The use of concentrations of 1 $mol\,dm^{-3}$ might be difficult if the compounds are not very soluble in water. However, no mark should be lost if the higher concentrations are suggested.

Student A

(b) $I_2(aq) + 2e^- \rightleftharpoons 2I^-(aq)$ $\qquad\qquad$ $E^{\ominus} = +0.54\,V$
$Fe^{3+}(aq) + e^- \rightleftharpoons Fe^{2+}(aq)$ $\qquad\qquad$ $E^{\ominus} = +0.77\,V$
So the possible overall reaction would be:
$2Fe^{3+}(aq) + 2I^-(aq) \rightarrow I_2(aq) + 2Fe^{2+}(aq)$
$E^{\ominus}$ for this is $0.77 - 0.54 = 0.23\,V$ so the reaction is possible.
In the reaction you would see iodine formed.
$Br_2(aq) + 2e^- \rightleftharpoons 2Br^-(aq)$ $\qquad\qquad$ $E^{\ominus} = +1.09\,V$
$Fe^{3+}(aq) + e^- \rightleftharpoons Fe^{2+}(aq)$ $\qquad\qquad$ $E^{\ominus} = +0.77\,V$
So the possible overall reaction would be:
$2Fe^{3+}(aq) + 2Br^-(aq) \rightarrow Br_2(aq) + 2Fe^{2+}(aq)$
$E^{\ominus}$ for this is $0.77 - 1.09 = -0.32\,V$ so the reaction is not possible.

Student B

(b) Possible equation is $2Fe^{3+}(aq) + 2I^-(aq) \rightarrow I_2(aq) + 2Fe^{2+}(aq)$

Overall $E^{\ominus} = 2 \times 0.77 - 0.54 = 1.00\,V$

So the reaction can take place and you will see the brown colour of iodine in solution and the yellow Fe^{3+} would go green.

Possible equation is $2Fe^{3+}(aq) + 2Br^-(aq) \rightarrow Br_2(aq) + 2Fe^{2+}(aq)$

Overall $E^{\ominus} = 2 \times 0.77 - 1.09 = +0.45\,V$

So the reaction can take place and you will see brown bromine in solution and the yellow Fe^{3+} would go green.

ⓔ Student A correctly predicts the outcome of the possible reactions and scores 4 marks. However, the statement that iodine is produced is just a statement and not a description of what would be seen so no marks are scored for that part.

Student B takes the trouble to give the correctly balanced overall equations but this leads to an error in interpreting the electrode potential data. It is incorrect to double the value of the electrode potential. They are used as given and no multiplication should be used to match the balancing numbers in the equation. The first prediction is still correct and the examiner would probably allow 2 marks for this but the second prediction is wrong and the student loses these marks. On the other hand the description of the reaction between $Fe^{3+}(aq)$ and $2I^-(aq)$ is correct, so Student B's overall mark is 3.

Student A

(c) $2I^-(aq) \rightarrow I_2(aq) + 2e^- \times 5$

$5e^- + MnO_4^-(aq) + 8H^+(aq) \rightarrow Mn^{2+} + 4H_2O(l) \times 2$

Overall equation is:

$10e^- + 2MnO_4^-(aq) + 16H^+(aq) + 10I^-(aq) \rightarrow 2Mn^{2+} + 5I_2(aq) + 4H_2O(l) + 10e^-$

Student B

(c) $2I^-(aq) \rightarrow I_2(aq) + 2e^-$

$5e^- + MnO_4^-(aq) + 8H^+(aq) \rightarrow Mn^{2+} + 4H_2O(l)$

which means:

$2MnO_4^-(aq) + 16H^+(aq) + 10I^-(aq) \rightarrow 2Mn^{2+} + 8H_2O(l) + 5I_2(aq)$

ⓔ In this part Student B does better than Student A and gets all 3 marks. Student A gets 1 mark but might be awarded 2 marks. The multiplication factor for each of the two half-equations has been identified correctly but there are two errors in the overall equation.

The $10e^-$ should not be included and, although the examiner might be prepared to accept this, the student has put $4H_2O(l)$ on the right-hand side of the equation instead of $8H_2O(l)$ and this means 1 mark is definitely lost. When combining half-equations it is easy to forget to multiply all the components by the necessary factor. It is always a good idea to have a good look at what you have done to try and avoid the error. Remember *both* symbols and charge have to balance in any equation.

ⓔ **Overall, Student A scores 11 or 12 out of 14 while Student B obtains 11.**

Question 8 **Fuel cells**

(a) **Explain the changes that take place at each electrode in a hydrogen–oxygen fuel cell with an acidic membrane.**
Give the overall reaction that is taking place in the cell. (4 marks)

(b) **Under standard conditions, a fuel cell can produce a voltage of 1.23 V.**
If the cell was used in a vehicle this voltage would be less than 1.23 V.
Suggest two reasons why the voltage might be less. (2 marks)

(c) **State three advantages that using fuel cells in vehicles might have over using petrol as the source of energy.** (3 marks)

(d) **Suggest two ways in which hydrogen might be stored in a vehicle using a fuel cell.** (2 marks)

Total: 11 marks

ⓔ In this question three different command words are employed. In (a), 'explain' requires a statement supported by appropriate equations; in (b) and (d) 'suggest' indicates that you have to use your knowledge and apply it to a problem that may not be explicitly on the specification; in (c), 'state' requires a simple statement with no supporting evidence.

Student A

(a) Hydrogen and oxygen are supplied to the fuel cell. At the cathode hydrogen gas is converted to hydrogen ions:
$H_2(g) \rightarrow 2H^+(aq) + 2e^-$
At the anode oxygen is converted to water by reaction with the hydrogen ions:
$\frac{1}{2}O_2(g) + 2H^+(aq) + 2e^- \rightarrow H_2O(l)$
Overall, the equation is:
$H_2(g) + O_2(g) \rightarrow H_2O(l)$

Student B

(a) $H_2(g) \rightarrow 2H^+(aq) + 2e^-$
$O_2(g) + 4H^+(aq) + 4e^- \rightarrow 2H_2O(l)$
$2H_2(g) + O_2(g) \rightarrow 2H_2O(l)$

ⓔ Both the students know the chemistry involved. Student A gets the 4 marks; Student B scores 3. All that Student B needed to do was mention that the gases are supplied to the cell externally. Having seen that this part carries 4 marks, it would have made sense to pause and try to think of a further point. The marks available usually give an idea of how many points the examiner is looking for.

Student A

(b) The pressure of the gas may mean the concentration of the gas is less.
The vehicle operates at a high temperature.

Student B

(b) The vehicle is not at standard conditions because the temperature is not standard. The rate of the reaction will vary depending how fast the vehicle is travelling.

ⓔ Student A has picked up two fairly straightforward marks. Student B is allowed 1 mark for the first statement but the second statement, while no doubt true, would not lead to a change in the voltage. This is a common confusion that is met on examination papers in a variety of ways. The energy change of a reaction is not affected by the rate at which it occurs. When using electrode potentials it is tempting to associate the size of the voltage with the rate at which the process will occur. A large electrode potential does not indicate a fast reaction and a reaction may occur quite readily even if the electrode potential is low.

Student A

(c) Fossil fuels will eventually run out whereas sources of hydrogen will not. Hydrogen is more efficient as a fuel. Hydrogen produces less pollution and no CO_2.

Student B

(c) Hydrogen is much lighter to carry. Hydrogen is cheap to produce. Hydrogen produces no pollution.

ⓔ Student A offers three good advantages; Student B is clearly guessing. Hydrogen cannot just be said to be lighter — it depends how it is held in the vehicle. Hydrogen is unlikely to be a cheaper alternative — it takes energy to produce it. It is unwise to suggest that hydrogen will produce no pollution — the source of oxygen will be the air, and the nitrogen it contains might still produce some polluting oxides. In addition, the manufacture of hydrogen needed for the fuel cells might produce some pollution. Student B fails to score.

In questions of this type there is no substitute for learning the content of the specification and being ready simply to give the answers suggested there. Student B might get some credit as the ideas are not wholly wrong, but it is silly to risk being caught out by mark schemes, which are always rigid.

Student A

(d) Adsorbed onto the surface of a solid. As a liquid under pressure.

Student B

(d) Absorbed into a solid carrier. Liquefied.

ⓔ Both students have learned this and both get 2 marks.

ⓔ **Overall, Student A scores the full 11 marks, while Student B gets 7, which is only grade-D standard. It is important to realise that exam papers are likely to contain some part of a question (or more) focused on the use of chemistry in everyday situations or where it is of international concern.**

Question 9 Transition metal chemistry and redox titrations

(a) (i) What is meant by the term transition element? (2 marks)

(ii) Complete the electron configuration of the iron atom:

$1s^2\ 2s^2\ 2p^6\ \ldots\ldots$ (1 mark)

(iii) Write the electron configuration of Fe^{2+} and Fe^{3+} ions. (2 marks)

(b) (i) Aqueous Fe^{2+} ions react with aqueous hydroxide ions. Write an ionic equation for this reaction and state what you would see. (2 marks)

(ii) The product formed in reaction between aqueous Fe^{2+} ions and aqueous hydroxide ions slowly darkens and eventually turns 'rusty'. What has happened to cause this colour change? (1 mark)

(c) The dichromate ion, $Cr_2O_7^{2-}$, is an oxidising agent that can be used in laboratory analysis. It reacts with acidified Fe^{2+} ions to form Cr^{3+} and Fe^{3+} ions:

$Cr_2O_7^{2-}(aq) + 14H^+(aq) + 6e^- \rightarrow 2Cr^{3+}(aq) + 7H_2O(l)$

$Fe^{2+}(aq) \rightarrow Fe^{3+}(aq) + e^-$

(i) Construct the full ionic equation for this reaction. (1 mark)

(ii) Calculate the volume of $0.0100\,mol\,dm^{-3}$ potassium dichromate required to react with $20.0\,cm^3$ of $0.0500\,mol\,dm^{-3}$ acidified iron(II) sulfate. (3 marks)

Total: 12 marks

ⓔ The command word 'construct' in part (c)(i) indicates that you must use the information given in the question to obtain the full ionic equation.

Student A

(a) (i) An element that forms one or more stable ions that have partly filled d-orbitals

Student B

(a) (i) An element that has partly filled d-orbitals

ⓔ Student A gets both marks but Student B scores only 1 mark. There are two key marking points: 'partly filled d-orbitals' is essential but it must also be clearly stated that the element forms one or more ions that have partly filled d-orbitals.

Student A

(a) (ii) $1s^2\ 2s^2\ 2p^6\ 3s^2\ 3p^6\ 3d^6\ 4s^2$

Student B

(a) (ii) $1s^2\ 2s^2\ 2p^6\ 3s^2\ 3p^6\ 4s^2\ 3d^6$

ⓔ Both students get the mark. The order of $3d$ and $4s$ is acceptable as either $3d$ followed by $4s$ or vice versa. However, if it is written as $4s^2 \, 3d^6$ mistakes might be made when it comes to writing electron configurations of ions.

Student A

(a) (iii) Fe^{2+}: $1s^2 \, 2s^2 \, 2p^6 \, 3s^2 \, 3p^6 \, 3d^6$
       Fe^{3+}: $1s^2 \, 2s^2 \, 2p^6 \, 3s^2 \, 3p^6 \, 3d^5$

Student B

(a) (iii) Fe^{3+}: $1s^2 \, 2s^2 \, 2p^6 \, 3s^2 \, 3p^6 \, 4s^2 \, 3d^4$
       Fe^{3+}: $1s^2 \, 2s^2 \, 2p^6 \, 3s^2 \, 3p^6 \, 4s^2 \, 3d^3$

ⓔ Student A gets both marks but Student B fails to score. The way in which Student B wrote the configuration for the Fe atom in (a)(ii) reflects the fact that the $4s$-subshell fills before the $3d$-subshells. However, the danger of writing it this way is that when ions are formed, the tendency is to remove electrons from the $3d$-orbitals first, when in fact the first electrons to be lost are always the outer electrons (in this case, the electrons in the $4s$-orbital).

Student A

(b) (i) $Fe^{2+}(aq) + 2OH^-(aq) \rightarrow Fe(OH)_2(s)$
       Green precipitate

Student B

(b) (i) $Fe^{2+} + 2OH^- \rightarrow Fe(OH)_2$
       $Fe(OH)_2$ is a blue–green gelatinous precipitate.

ⓔ Both get 2 marks although Student B ought to include state symbols.

Student A

(b) (ii) Fe^{3+} is more stable than Fe^{2+} and therefore iron(ɪɪ) compounds are readily oxidised. The rust colour is due to $Fe(OH)_3$ being formed.

Student B

(b) (ii) $Fe(OH)_3$ is a rust–brown gelatinous precipitate and therefore $Fe(OH)_2$ must have changed to $Fe(OH)_3$.

ⓔ Both students get the mark.

Student A

(c) (i) $Cr_2O_7{}^{2-} + 14H^+ + 6Fe^{2+} \rightarrow 6Fe^{3+} + 2Cr^{3+} + 7H_2O$

Student B

(c) (i) $Cr_2O_7{}^{2-} + 14H^+ + 5e^- + Fe^{2+} \rightarrow Fe^{3+} + 2Cr^{3+} + 7H_2O$

ⓔ Student A gets the mark but Student B does not. The full ionic equation does not contain any electrons. It is essential to scale the two half-ionic equations so that the electrons cancel out.

Student A

(c) (ii) mols of Fe^{2+} = 0.0500 × 20.0/1000 = 0.001 mol
mols of $Cr_2O_7^{2-}$ = 6 × moles of Fe^{2+} = 0.001 × 6 = 0.006 mol
1 dm^3 contains 0.01 mol of $Cr_2O_7^{2-}$
so volume of $Cr_2O_7^{2-}$ = 0.006/0.01 = 0.6 = 600 cm^3.

Student B

(c) (ii) moles of Fe^{2+} = moles of $Cr_2O_7^{2-}$ = 0.0500 × 20.0/100 = 0.001 mol
$n = cV/V = n/c$ = 0.001/0.01 = 0.1 dm^3 = 100 cm^3.

ⓔ Both students get 2 marks but for different reasons. Student A has correctly calculated the moles of Fe^{2+} but has multiplied this value by 6 instead of dividing it by 6. Student B has calculated the moles of Fe^{2+} correctly, but he/she has not used the 1:6 molar ratio. Both students have gone on to calculate the volume of $Cr_2O_7^{2-}$ required. Both display good examination technique by showing their working. Titrations calculations are usually designed so that the volume added from the burette is around 25 cm^3. Calculations rarely involve volumes over 50 cm^3. The answers obtained by both students should have prompted them to check their calculations for errors. The correct value is 16.7 cm^3.

ⓔ **Overall, Student A scores 11 out of 12 and Student B scores 7.**

Question 10 Transition metal chemistry

(a) Explain the meaning of the terms 'ligand' and 'coordinate bond'. (2 marks)

(b) Stereoisomerism is sometimes shown by transition metal complex ions.
Using a suitable named example in each case, show how transition metal complex ions can form:
(i) *cis–trans* isomers
(ii) optical isomers (8 marks)

(c) Ligand exchange can often occur when transition metal complex ions react. These will take place if a new complex ion can be formed that has a greater stability. An example of a ligand exchange reaction is the formation of $[CoCl_4]^{2-}$ from $[Co(H_2O)_6]^{2-}$.
(i) The stability of a complex ion may be expressed by providing its stability constant. Define the stability constant of $[CoCl_4]^{2-}$. (2 marks)
(ii) Describe how you would convert $[Co(H_2O)_6]^{2-}$ into $[CoCl_4]^{2-}$ and what you would see as the reaction took place. (3 marks)

Total: 15 marks

ⓔ Part (b) is a free-response question and requires careful planning. There are 8 marks available so it is likely that eight different points have to be made. The question is divided into parts (i) and (ii) so it is likely that 4 marks will be allocated to each part. Each type of isomerism is best illustrated by suitably labelled diagrams.

Student A

(a) A ligand is a lone pair donor. A coordinate bond is the same as a dative bond.

Student B

(a) A ligand is a lone pair donor. A coordinate bond is formed between the ligand and the vacant *d*-orbitals of the Fe^{2+}. The ligand supplies both electrons for the bond.

ⓔ Student A scores only 1 mark as the definition of a coordinate bond is insufficient. Student B gets both marks as detail is provided in the explanation of coordinate bond.

Student A

(b) (i) *cis–trans* isomerism occurs when a complex has two different ligands. In one version the ligands of one type are alongside each other while in the other version they are opposite. An example is the dichloro-diamminonickel complex shown below.

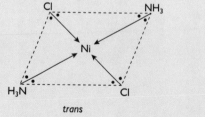

trans

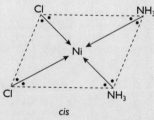

cis

OCR(A) A2 Chemistry

(ii) Optical isomerism occurs when one stereoisomer is a reflection of the other. The ligands must be bidentate. An example is with nickel. The ligand is shown as a double-headed arrow in my diagram.

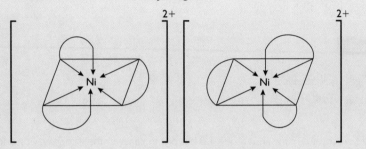

As you can see the second version is a reflection of the first and it is non-superimposable.

(b) **(i)** Nickel can form a complex ion containing ammonia and chloride. In the one version of the complex the ammonia molecules are next to each other and so are the chloride ions. This is called *cis*. But also they might be opposite each other across the complex ion and then it would be *trans*.

(ii) Optical isomers occur when you have ligands that attach at two points on the metal ion. Nickel forms one of these as well. If you have three of them they can be attached and then there are two versions that are mirror images of each other.

ⓔ The marking points for this question are:
- *cis*–*trans*: named example, give shape, explanation of difference between *cis*–*trans*, identify *cis* and *trans* correctly
- optical: named example, explain multidentate ligand required, explain isomers are reflections of each other, indicates how the ligands are attached

Student A has done well and has made a good attempt to explain the two types of isomerism. The use of clear diagrams earns almost all the marks. For *cis*–*trans* all the marks could be obtained from a clearly labelled diagram and for optical isomerism all but 1 mark is possible. Student A gets 7 of the 8 marks in this way but an actual example of a complex ion showing optical isomerism is not given, which loses 1 mark.

Student B may know the answer but tries to answer the question without using diagrams. This is almost impossible. Perhaps the third and fourth marks can be awarded for the *cis*–*trans* explanation and the second and third for the optical isomerism, giving a total of 4 marks.

An exam question may ask you to draw diagrams to illustrate your answer. However, even if the question does not say so, you can always use diagrams if you find it hard to put into words what you want to say.

Student A

(c) (i) $K = \dfrac{[CoCl_4^{2-}]}{[Co^{2+}][Cl^-]^4}$

Student B

(c) (i) $K_{stab} = \dfrac{[Co^{2+}][Cl^-]^4}{[CoCl_4^{2-}]}$

ⓔ Student A scores 2 marks, but Student B has put the answer the wrong way up and only gets 1 mark. In each case it would have been better, and safer, to write $[Co^{2+}]$ as $[Co(H_2O)_6^{2+}]$.

Student A

(c) (ii) The addition of chloride ions will convert $[Co(H_2O)_6]^{2+}$ into $[CoCl_4]^{2-}$. You would see the pink solution of $[Co(H_2O)_6]^{2+}$ change to a blue colour showing that $[CoCl_4]^{2-}$ had been made.

Student B

(c) (ii) Excess concentrated hydrochloric acid must be added to the water complex ion. When it is added the colour change that occurs is from pink to blue.

ⓔ Student B obtains all 3 marks, but Student A earns only 2 marks because he/she leaves out the excess chloride ions.

ⓔ **Student A has done well scoring 12 out of 15 marks. Student B obtains 10 marks.**

Question 11 **Synoptic question**

Copper reacts with nitric acid, HNO_3, but the products of the reaction depend on the concentration of the acid.

If the acid is dilute then the following reaction takes place:

copper + dilute nitric acid

$Cu(s) \rightarrow Cu^{2+}(aq) + 2e^-$

$4H^+(aq) + NO_3^-(aq) + 3e^- \rightarrow NO(g) + 2H_2O(l)$

If the nitric acid is concentrated then the reaction is:

copper + concentrated nitric acid

$Cu(s) \rightarrow Cu^{2+}(aq) + 2e^-$

$2H^+(aq) + NO_3^-(aq) + e^- \rightarrow NO_2(g) + H_2O(l)$

(a) Write balanced equations for each of the above reactions.   (5 marks)

(b) In an experiment, some nitric acid is reacted with 1.27 g of copper and it is found that 320 cm³ of gas is produced. Deduce whether the acid used in this experiment was dilute or concentrated.
Show *all* your working.   (4 marks)

                                                                    Total: 9 marks

ℯ Part (a) is fairly straightforward — you have to use the half-equations provided to determine the overall ionic equations. Part (b) is much more difficult; you have to relate the moles of Cu and the moles of gas to your full ionic equations. The command words in this question are deliberately vague.

Student A

(a) The electrons have to balance, hence for dilute acid
$Cu(s) \rightarrow Cu^{2+}(aq) + 2e^-$; multiply by 3
$4H^+(aq) + NO_3^-(aq) + 3e^- \rightarrow NO(g) + 2H_2O(l)$; multiply by 2
$3Cu(s) \rightarrow 3Cu^{2+}(aq) + 6e^-$
$8H^+(aq) + 2NO_3^-(aq) + 6e^- \rightarrow 2NO(g) + 4H_2O(l)$
Balanced equation is:
$8H^+(aq) + 2NO_3^-(aq) + 3Cu(s) \rightarrow 3Cu^{2+}(aq) + 2NO(g) + 4H_2O(l)$
For concentrated acid
$Cu(s) \rightarrow Cu^{2+}(aq) + 2e^-$
$2H^+(aq) + NO_3^-(aq) + 1e^- \rightarrow NO_2(g) + H_2O(l)$ multiply by 2
$Cu(s) \rightarrow Cu^{2+}(aq) + 2e^-$
$4H^+(aq) + 2NO_3^-(aq) + 2e^- \rightarrow NO_2(g) + H_2O(l)$
Balanced equation is:
$4H^+(aq) + 2NO_3^-(aq) + Cu(s) \rightarrow Cu^{2+}(aq) + NO_2(g) + H_2O(l)$

Student B

(a) Dilute nitric:

$8H^+(aq) + 2NO_3^-(aq) + 3Cu(s) \rightarrow 3Cu^{2+}(aq) + 2NO(g) + 4H_2O(l)$

Conc. nitric: $4H^+(aq) + 2NO_3^-(aq) + Cu(s) \rightarrow Cu^{2+}(aq) + 2NO_2(g) + 2H_2O(l)$

ⓔ The marking points for this question are:
- dilute acid — multiply copper half-equation by 3, multiply acid half-equation by 2, correct balanced equation
- concentrated acid — multiply acid half-equation by 2, correct balanced equation

Student A gets all 3 marks for the equations for the dilute acid but carelessly loses 2 marks for the concentrated acid equations by failing to multiply the product side of the equation by 2.

Student B poses a dilemma for the examiner. Clearly, he/she is able and has correctly deduced the balanced equations for both reactions, but not all of the working has been shown. Student B would probably get 4 of the 5 marks.

Student A

(b) With dilute nitric acid, 3 mol of Cu will produce 2 mol of NO(g).
With concentrated nitric acid, 1 mol of Cu will produce 1 mol of NO_2(g).
The amount in moles of copper reacted is $1.27/63.5 = 0.02$.
The amount in moles of gas produced is $320/24\,000 = 0.0133$.
Mole ratio Cu:gas

0.02:0.0133

1.5:1

∴ 3:2

3 mol of Cu produce 2 mol of gas and therefore the acid must be dilute.

Student B

(b) moles of Cu used = 0.02

moles of gas = $320/24 = 13.3$

ⓔ The marking points for this question are moles of copper used, moles of gas produced, molar ratio Cu:gas, relates molar ratio to correct equation.

Student A scores the final 4 marks by deducing correctly that the dilute acid was used.

Student B scores 1 mark for the moles of copper used but scores no more marks. In the final part of the calculation Student B has incorrectly divided $320\,cm^3$ by $24\,dm^3$ and this has given a value that does not relate to either equation and Student B is, therefore, unable to continue with the calculation.

ⓔ **Overall, Student A scores 7 out of 9 marks even though one of the equations is wrong. Student B, who correctly gives both equations, only scores 5 out of 9 marks.**

Knowledge check answers

1 **(a)** rate would double
   (b) rate would decrease by factor of 4 (quartered) $(\frac{1}{2})^2 = \frac{1}{4}$
   (c) rate would increase by factor of $3 \times 3^2 = 27$ times faster

2 **(a)** units of k are $dm^6 mol^{-2} s^{-1}$ or $mol^{-2} dm^6 s^{-1}$
   (b) units of k are $dm^{12} mol^{-4} s^{-1}$ or $mol^{-4} dm^{12} s^{-1}$

3 Using experiments 1 and 2: concentration of $[BrO_3^-]$ is doubled and the rate doubles. Therefore the reaction is first order with respect to $[BrO_3^-]$.
   Using experiments 2 and 3: concentration of $[Br^-]$ is doubled and the rate doubles. Therefore the reaction is first order with respect to $[Br^-]$.
   The units of the rate constant show that the overall reaction is fourth order. Therefore the reaction is second order with respect to $[H^+]$

4 slow step $ICl + H_2 \rightarrow HCl + HI$
   fast step $HI + ICl \rightarrow HCl + I_2$
   The two steps add up to give the overall equation:
   $2ICl + H_2 \rightarrow 2HCl + I_2$

5 Low temperature would favour the forward exothermic reaction, so the equilibrium would move to the right-hand side.
   High pressure would move the equilibrium position to the right because there are fewer moles of gas on the right-hand side.

6 **(a)** units $= dm^3 mol^{-1}$ (or $mol^{-1} dm^3$)
   (b) units $= dm^3 mol^{-1}$ (or $mol^{-1} dm^3$)

7 Number of moles of each gas must be divided by the volume ($4 dm^3$) in order to get the concentrations of each gas:
   $[N_2] = 1.25$, $[H_2] = 2.5$, $[NH_3] = 1.25$
   $K_c = 1.25^2/(1.25 \times 2.5^3) = 0.08 dm^6 mol^{-2}$

8 There are an equal number of moles of gas on each side of the equation. Therefore K_c does not have any units and it does not matter whether you use the number of moles or the concentration of each component — the answer will be the same.
   number of moles of each gas at equilibrium: $H_2(g) = 0.3$, $I_2(g) = 0.2$, $HI(g) = 0.2$
   concentration of each gas at equilibrium: $[H_2(g)] = 0.15$, $[I_2(g)] = 0.1$, $[HI(g)] = 0.1$
   using moles: $K_c = 0.2^2/(0.3 \times 0.2) = 0.67$
   using concentrations: $K_c = 0.1^2/(0.15 \times 0.1) = 0.67$

9 **(a)** $CH_3COO^-K^+$
   (b) $(HCOO^-)_2 Mg^{2+}$
   (c) $Ca^{2+}_3(PO_4^{3-})_2$ — (could also form $Ca^{2+}HPO_4^{2-}$ or $Ca^{2+}(H_2PO_4^-)_2$

10 **(a)** OH^-
   (b) SO_4^{2-}
   (c) NH_2^-
   (d) $C_6H_5COO^-$

11 **(a)** H_3O^+
   (b) H_2SO_4
   (c) NH_4^+
   (d) $CH_3NH_3^+$

12 $CH_3CO_2H + HNO_3 \rightleftharpoons CH_3CO_2H_2^+ + NO_3^-$

13 **(a)** $pH = 1.0$
   (b) $pH = 2.0$
   (c) $pH = 3.0$

14 B, A, C.

15 $6.3 \times 10^{-5}/(1.7 \times 10^{-5}) = 3.7$ times stronger

16 As temperature increases so does K_w, therefore so does $[H^+(aq)]$ and $[OH^-(aq)]$, hence increasing temperature moves the equilibrium to the right indicating that the forward reaction is endothermic and ΔH will therefore be positive.

17 **(a)** $pH = 0.82$
   (b) $pH = 5.07$
   (c) $pH = 13.18$
   (d) moles of unreacted $HCl = 0.01$
       concentration of unreacted $HCl = 0.333 mol dm^3$
       pH of solution $= 0.48$

18 If HIn represents the formula of bromocresol green, then $K_{In} = [H^+][In^-]/[HIn]$. $[In^-]$ is blue and $[HIn]$ is yellow. The green colour will be formed when $[In^-] = [HIn]$ such that $K_{In} = [H^+]$. The pH when the green colour is formed is 4.7.

19 **(a)** $\frac{1}{2}Cl_2(g) \rightarrow Cl(g)$
   (b) $Ca^{2+}(g) + 2Cl^-(g) \rightarrow CaCl_2(s)$
   (c) $Mg^+(g) \rightarrow Mg^{2+}(g) + 1e^-$

20 **(a)** $MgBr_2$, $NaBr$, KBr
   (b) CaF_2, $CaCl_2$, $BaCl_2$, $BaBr_2$

21 **(a)** $Ca^{2+}(g) + aq \rightarrow Ca^{2+}(aq)$
   (b) $Ca(OH)_2(s) + aq \rightarrow Ca^{2+}(aq) + 2OH^-(aq)$
   (c) $Ca^{2+}(g) + 2OH^-(g) \rightarrow Ca(OH)_2(s)$

22 **(a)** ΔS will be negative (the movement of particles in ice is more restricted)
   (b) ΔS will be positive (the particles in an aqueous solution have more freedom to move than particles in the solid)
   (c) ΔS will be negative (the oxygen molecules have reacted and are not free to move)
   (d) ΔS will be negative (the reduction in overall volume as the reaction takes place reduces the freedom to move)

23 The equation for the reaction is $2Na(s) + \frac{1}{2}O_2(g) \rightarrow Na_2O(s)$
   The entropy change, ΔS, is $72.8 - (2 \times 51.0 + \frac{1}{2} \times 102.5) = -80.5 J mol^{-1} K^{-1}$

24 **(a)** $MnO_4^-(aq) + 8H^+(aq) + 5V^{2+}(aq) \rightarrow 5V^{3+}(aq) + Mn^{2+}(aq) + 4H_2O(l)$
   (b) $3MnO_4^-(aq) + 24H^+(aq) + 5V^{2+}(aq) + 15H_2O(l) \rightarrow 5VO_3^-(aq) + 30H^+(aq) + 3Mn^{2+}(aq) + 12H_2O(l)$
       This is simplified by cancelling the H^+ and H_2O that appear on both sides of the equation to give:
       $3MnO_4^-(aq) + 5V^{2+}(aq) + 3H_2O(l) \rightarrow 5VO_3^-(aq) + 6H^+(aq) + 3Mn^{2+}(aq)$
   (c) $Cr_2O_7^{2-}(aq) + 14H^+(aq) + 3SO_2(aq) + 6H_2O(l) \rightarrow 3SO_4^{2-}(aq) + 12H^+(aq) + 2Cr^{3+}(aq) + 7H_2O(l)$
       This can then be simplified to:
       $Cr_2O_7^{2-}(aq) + 2H^+(aq) + 3SO_2(aq) \rightarrow 3SO_4^{2-}(aq) + 2Cr^{3+}(aq) + H_2O(l)$
   (d) $8H^+(aq) + 2NO_3^-(aq) + 3Cu(s) \rightarrow 3Cu^{2+}(aq) + 2NO(g) + 4H_2O(l)$

25 (a) $Mg(s) \rightarrow Mg^{2+}(aq) + 2e^-$ $E^\ominus = +2.37V$
$Zn^{2+}(aq) + 2e^- \rightarrow Zn(s)$ $E^\ominus = -0.76V$
Therefore the overall cell potential is 1.61 V

(b) $Fe^{3+}(aq) + e^- \rightarrow Fe^{2+}(aq)$ $E^\ominus = +0.77V$
$Sn^{2+}(aq) \rightarrow Sn^{4+}(aq) + 2e^-$ $E^\ominus = -0.15V$
Therefore the overall cell potential is 0.62 V

(c) $Br_2(aq) + 2e^- \rightarrow 2Br^-(aq)$ $E^\ominus = +1.09V$
$2I^-(aq) \rightarrow I_2(aq) + 2e^-$ $E^\ominus = -0.54V$
Therefore the overall cell potential is 0.55 V

(d) $Zn(s) \rightarrow Zn^{2+}(aq) + 2e^-$ $E^\ominus = +0.76V$
$I_2(aq) + 2e^- \rightarrow 2I^-(aq)$ $E^\ominus = +0.54V$
Therefore the overall cell potential is 1.30V

(e) $Br_2(aq) + 2e^- \rightarrow 2Br^-(aq)$ $E^\ominus = +1.09V$
$Sn^{2+}(aq) \rightarrow Sn^{4+}(aq) + 2e^-$ $E^\ominus = -0.15V$
Therefore the overall cell potential is 0.94V

26 Equation at oxygen electrode: $1\frac{1}{2}O_2(g) + 6H^+(aq) + 6e^- \rightarrow 3H_2O(l)$
Equation at the fuel electrode: $CH_3OH(l) + H_2O \rightarrow 6H^+(aq) + 6e^- + CO_2(g)$
Overall equation: $CH_3OH(l) + 1\frac{1}{2}O_2(g) \rightarrow CO_2(g) + 2H_2O(l)$

27 (a) +2
(b) +2
(c) +3
(d) +2
(e) +3

28 (a) 6
(b) 2
(c) 6
(d) 6
(e) 4

29

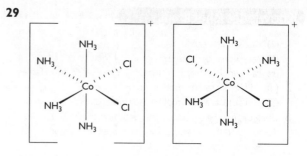

Shape is octahedral, bond angles are 90°, the oxidation number of the Co is +3 and the type of isomerism is *cis – trans*

$$en = H_2NCH_2CH_2NH_2$$

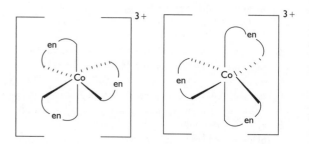

Shape is octahedral, bond angles are 90°, the oxidation number of the Co is +3 and the type of isomerism is optical isomerism

30 (a) $4NH_3 + [Cu(H_2O)_6]^{2+} \rightarrow [Cu(NH_3)_4(H_2O)_2]^{2+} + 4H_2O$
(b) $Cl^- + [Co(NH_3)_6]^{3+} \rightarrow [Co(NH_3)_5Cl]^{2+} + NH_3$
(c) $SCN^- + [Fe(H_2O)_6]^{3+} \rightarrow [Fe(H_2O)_5(SCN)]^{2+} + H_2O$

31 moles of $S_2O_3^{2-}(aq) = 1.80 \times 10^{-3} =$ moles of Cu^{2+}
mass of $Cu^{2+} = 1.80 \times 10^{-3} \times 63.5 = 0.1143g$
%Cu $= (0.1143/0.426) \times 100 = 26.8\%$ (to 3 s. f.)

Note: **bold** page numbers indicate defined terms.

A

acid dissociation constant   21–22, 23, 27

acids and bases

Brønsted–Lowry   20

buffer solutions   26–27

pH calculations   23–24

pH changes and indicators   28–30

questions & answers   76–78

strengths of   21–22, 30–31

water, ionic product   25

activation energy (E_a)   8, 9, 10, 33

average bond enthalpy   32, 33

B

bases *see* acids and bases

batteries (storage cells)   49–50

blood, control of pH in   28

Boltzmann distribution of energies   9, 10

bond dissociation enthalpy   35

Born–Haber cycles   34–35

questions & answers   79–84

Brønsted–Lowry acids and bases   20

buffer solutions   **26–27**

everyday uses of   28

C

catalysts

and equilibrium   17

and rate of reaction   10

transition metals   53–54

collision theory of reactivity   8

complete dissociation   23

complex ions   **54–55**

ligand substitution of   56–57

stability constant   57–58

stereoisomerism in   55–56

concentration

and equilibrium position   17

feasibility of reactions   49

hydrogen ions, pH scale   22–24

initial rates method   14

and K_c (equilibrium constant)   17–19

and order of reaction   11–13

and rate of reaction   9, 11

concentration–time graphs   10–11, 12, 14

conjugate acids   **20**

conjugate bases   **20**

in buffer solutions   26–27, 28

conjugate pairs   **20**

coordination number   **55**

D

dynamic equilibrium   16

E

electrode potentials   46–49

questions & answers   88–90

endothermic reactions   16, 18, 32

energy

activation energy (E_a)   8, 9, 10, 33

Boltzmann distribution   9, 10

electrode potentials   46–49

enthalpy   32–40

entropy   40–43

fuel cells   50–52

enthalpy changes   32–34

see also lattice enthalpies

definitions for standard   35–36

hydration and solution   39–40

enthalpy of hydration   **39**

enthalpy of neutralisation   **31**

enthalpy of solution   **39–40**

entropy   40–43, 85–87

equilibrium

dynamic   16

factors affecting   16–17

questions & answers   69–71

temperature   42–43

equilibrium constant (K_c)   17–19

and stability constant (K_{stab})   57–58

equilibrium law   17

exothermic reactions   16, 18, 32

F

feasibility of reactions

effect of concentration   49

and free energy   42

first electron affinity   **36**, 37

first ionisation energy   **36**, 37

free energy   42

calculating changes in   43

questions & answers   85–87

fuel cells   50–52

questions & answers   91–92

G

Gibbs free energy *see* free energy

H

half-life   12

Hess's law   33–34, 35, 37, 40

hydration, enthalpy of   39

I

indicators   **28–29**

choosing   29–30

ionic equations   43–44

ionic product of water   25

iron(II)–manganate(VII) reaction   58–59

isomerism   55–56, 96–97

K

K_a (acid dissociation
constant) 21–22
pH calculations 23–24, 27
K_c (equilibrium constant) 17–19
K_w (ionic product of water) 25–26

L

lattice enthalpies **34**
Born–Haber cycle 34–35
calculation of 36–38
factors affecting size of 38–39
le Chatelier's principle 16,
17, 18
ligands **54**
bidentate 55
polydentate 56
questions & answers 96–97
substitution reactions 56–57

N

neutralisation, enthalpy of 31

O

optical isomerism 56, 97
orders of reaction 11
determining from graphs 11–13
with respect to each
reagent 14
oxidation number 44–46, 58
oxidation states, transition
elements 53, 58

P

pH 22–23
of acids 23–24
of blood, control of 28
buffer solutions 26–28
and indicators 28–30
questions & answers 72–75
of strong alkalis 26
of water 25
pK_a 24
precipitation reactions 54
pressure
and equilibrium position 16, 18
standard 32

R

rate–concentration graphs 12–13
rate constant (k) 11
units of 15
rate-determining step 15–16
rate equation 11, 14–16
questions & answers 65–68
rate of reactions 8
and catalysts 10
concentration 9
measuring from graphs 10–11
measuring using initial rates
method 14
and temperature 9–10
reaction mechanisms, steps
in 15–16
redox reactions 44–46
transition elements 58–61
reversible reactions **16**
revision techniques 5–7

S

solubility, enthalpy of 39–40
stability constant (K_{stab}) 57–58
standard cell potential **46**
standard electrode potentials **46**
measuring 47–48
standard enthalpy change of
atomisation **35**
standard enthalpy change of
combustion 32, 33
standard enthalpy change of
formation 32, 34, **35**
standard enthalpy changes 32–33,
35–37, 39–40
storage cells 49–50
study skills 5–7
synoptic assessment 8, 99–100

T

temperature
and equilibrium position 16, 18
finding equilibrium 42–43
and rate of reaction 9–10
standard 32
terms used in unit
test 63–64
time management 6–7
titration curves 29–30
titrations 59–61
transition elements **52**
complexes 54–57
electron configurations 52–53
precipitation reactions 54
properties 53–54
questions & answers 93–98
redox reactions 58–59
redox titrations 59–61
stability constants 57–58

U

units
acid dissociation constant 22
ΔH and ΔG 43
equilibrium constant 18
rate constant 15
reaction rate 10

W

water, ionic product of 25